Ethical Issues in Human Cloning

Ethical Issues in Human Cloning

Cross-Disciplinary Perspectives

Edited by
Michael C. Brannigan
Center for the Study of Ethics
La Roche College

Seven Bridges Press, LLC
135 Fifth Avenue, New York, NY 10010-7101

Publisher: Ted Bolen
Managing Editor: Katharine Miller
Cover Design: Stefan Killen Design
Composition: Bytheway Publishing Services
Printing and Binding: Victor Graphics, Inc.

Library of Congress Cataloging-in-Publication Data
Ethical issues in human cloning : cross-disciplinary perspectives / edited by Michael C. Brannigan.
p. cm.
ISBN 1-889119-11-3 (pbk.)
1. Human cloning—Moral and ethical aspects.
I. Brannigan, Michael C., date.

QH442.2 .E845 2000
174'.25—dc21 00-009935
CIP

Manufactured in the United States of America
10 9 8 7 6 5 4 3 2 1

To my precious sisters
Maggie and Marie
whose hearts' dance
embraces us

Contents

Acknowledgments

A BOOK ON human cloning unavoidably raises questions of lineage and originality. And when it comes to tracing the lineage of a text, we discover that there may be one writer, but many authors. What is written owes its existence to the swell of characters and memories that give light to each page. This book is no exception. It is in essence an extension of the unique imprints of others, whose own originality is as indisputable as their fingerprints.

To begin with, I continue to owe an immense measure of thanksgiving to my wife Brooke, who, as my partner, bears the brunt of my moods, and yet never fails to offer her encouragement and insight. By the same token, though my entire family has had to endure my frequent absences, they have always remained present for me. And my steadfast friends, particularly my lifelong companions Jim Fanning, Tad Connerton, Ron Brooks, and Cliff Muldoon, have also patiently tolerated my silence.

My gratitude also extends to those colleagues who have encouraged me in their singular ways. Members of the Humanities Division deserve special praise. And I am indebted to Ed Brett, Astrid Kersten, and Barbara Coyne for their passionate commitment to scholarship. Our exceptional library staff has generously addressed my many inquiries and helped in procuring sources. We are indeed fortunate to have as our Library Director Cole Puvogel, who exudes the rare combination of professionalism and humaneness. Cindy Speer has been a most able and thorough assistant as well as a blessing in procuring permissions. Laverne Collins, Darlene Veghts, Grace Voytosh, and Sally Knapp continue to offer their special assistance. Susan Klimcheck and Joanne Ferrill always give their graceful support. And Leslie Kirby has been extremely helpful with her careful and patient reading of the manuscript.

Our Center for the Study of Ethics is in its third year. I offer my thanks to all those, both faculty and staff, who have supported our center, especially to our president, Monsignor William Kerr, for bringing the center out of gestation. And a special chorus of thanks to our former Academic Dean, now deceased, Regina Borum, who always provided gracious and unflinching support for the center and its aims, and to the center's staff members: administrator Janine

Bayer, Rose Marie Hogan, who has regularly volunteered her time to enhance the work of the center, Shane Campbell, for his creative guidance in our journal, and our hard-working intern, Darrah Price. Also, resounding thanks to the many students who have volunteered their time and energies to help out during our conferences and presentations.

Final chords of gratitude go to my publisher, Ted Bolen, for his dependability, unending support, and fresh optimism. I also thank Clay Glad for his editorial insights. All of the above comprise the lineage of this book.

Introduction: The Ethical Challenge

JUST ABOUT EVERYONE has heard of "Dolly," the lamb cloned from an udder cell of an adult sheep. This happened at the Roslin Institute, formerly known as the Animal Breeding Research Station, near Edinburgh, Scotland. Udder cells had been taken from the six-year-old ewe and frozen. As to the fate of this ewe, no one seems to know, although it's safe to guess that she may have ended up on the plates of some unknowing Scots.

So instead, let's talk about my sisters, Margaret and Marie, who, as natural identical twins, are about as close as possible to being pure clones. They share the same genotype, and they look remarkably alike. They even share a passion for dancing. Naturally gifted, they danced their way together from Newport, Rhode Island, to Stephens College in Missouri, becoming part of its faculty; now they run their own successful studio. At the same time, they are decidedly different in numerous ways. They are worlds apart in disposition and personality. Even in the domain of dance, they have different specialties—the temperament for ballet is radically unlike that for modern jazz.

The moral of this tale is clear. Even with natural clones such as my sisters, precise genetic copies do not translate into exact duplicates. In contrast to science fiction, science fact indicates that exact copies of Hitler, Mozart, and so on are impossible. What is possible has been demonstrated to us through the birth of Dolly: just as we had formerly produced genetic copies of molecules, cells, and plants, we could now produce genetic copies of animals. The more stunning news, however, is that we can now use adult cells as donors, cells that have already become specialized and differentiated.

Let's get back to Dolly. Dolly was born on July 5, 1996, and soon after the headlines first appeared—nearly eight months later, in February 1997—describing the successful cloning by Ian Wilmut and his colleagues, Dolly and Wilmut became overnight celebrities. There was the usual media frenzy and the ruminations of countless pundits.

The event, however, was not some overnight miracle. It was the result of long, arduous, painstaking labor. Wilmut, whose principal goal was to produce life-saving drugs for conditions such as cystic fibrosis by genetically altering sheep, rode on the shoulders of earlier researchers. Hans Spemann, in the early 1900s, theorized that transferring a nucleus from an embryo cell to an enucleated egg (an egg whose nucleus had been removed) would result in an entity with the same genetic make-up as the transferred nucleus. In other words, he submitted the possibility of cloning. In 1951, Robert Briggs and Thomas King actualized Spemann's theory by successfully cloning frogs. They did not use cells from adult frogs, however, for doing so was considered impossible; the cells would be fully differentiated, and it was the consensus among scientists that the more specialized the cells were, the less likelihood of cloning.

In the late 1960s, John Gurdon claimed to be able to clone frogs by transferring cells from their intestinal lining, that is, from specialized and differentiated cells. Whether or not his claim had merit, Gurdon unleashed the possibility that clones might result from adult, differentiated cells. In 1979, Karl Illmensee claimed to have cloned mice, making this the first cloning of mammals. With the weight of the evidence against him, the jury is still out as to whether or not his claim was valid, but even if his claim was true, the cloning still resulted from mice embryo cells and not from adult cells. And then came Steen Willadsen, who cloned a sheep, but still from an undifferentiated, embryo cell.

The challenge remained: Was it possible to clone an adult cell? The possibility appeared more feasible in July 1995, when Ian Wilmut and Keith Campbell produced a pair of lambs, Megan and Morag, from embryo cells that were starting to differentiate. Then Dolly's birth the next year sent out global shock waves. Not only did it prove that adult donor cells could be used for cloning, but it raised the specter that cloning could feasibly be applied someday to humans.

When Dolly's birth was announced, countries throughout the world had already initiated efforts to prohibit human cloning. Australia, Denmark, England, Germany, and Spain are among the countries outlawing human cloning. Opposition came from other groups, including the World Health Organization, numerous religious bodies such as the Vatican, and even the American Society of Reproductive Medicine. President Clinton immediately placed a moratorium on any federally funded cloning research, and he requested the National Bioethics Advisory Commission (NBAC) to investigate the matter further and to make appropriate recommendations. In June 1997, after hearing assorted testimonies, the panel advised legislation that would make human cloning illegal. As we see in this volume, not all scholars agreed with that conclusion.

It is difficult to be evenhanded when discussing human cloning. Is it the "dark side" of reproductive technologies? Or does it represent a golden opportunity for human enhancement? To be sure, for an entire planet standing at a critical crossroads, cloning offers both promises and perils. In agriculture, we herald the idea of reproducing prize cows for milking. Ecologically, animal cloning may enable the survival of near-extinct species. For humans, the promises extend into all sorts of possibilities, such as finding drugs that would alleviate diseases like cystic fibrosis, cultivating one's own bone marrow as well as solid organs for transplantation, and genetically altering animals such as pigs in order to provide perfectly compatible organs for transplantation into humans. As an extension of reproductive techniques, the possibilities in human cloning promise ways both to relieve infertility and to prevent the transmission of genetic diseases.

There are also disturbing possibilities, particularly when we consider what is traditionally regarded as the nucleus of society—the family, for which enough radical changes have already taken place in the past century. As we move into the twenty-first century, human cloning may pose the ultimate challenge to our notions of family, and its possibilities pose special hazards because the field of reproductive technology is without any real government regulation or oversight. Although there is currently a ban on federal funding of human cloning–related research, such research will no doubt attract private funding. And extreme caution will be needed to prevent the kind of profiteering that human cloning may engender. Moreover, since some scientists are no doubt already researching diligently the possibility of human cloning, there is probably more going on than we now know. Will the first human clones be kept secret for some time, especially if they are anomalies?

The following essays reflect four complementary perspectives: scientific, religious, philosophical, and legislative. They are written by the foremost scholars in their fields, whose engaging and provocative viewpoints offer a broad spectrum of positions, ranging from outright condemnation of human cloning to more favorable stances. Yet even the most favorable positions are laced with a cautious optimism and the realization that, in this new world of reproductive technologies, we need, more than ever, to proceed with prudence.

Indeed, human cloning profoundly challenges our deepest and most cherished beliefs about what it means to be human. Over the entrance to Plato's Academy, the first "university" in the Western world, were the words *Know thyself,* and this mandate has remained the quintessential motto of Western philosophy. Without a doubt, human cloning compels us to dig more deeply into its significance. Who, or what, in essence are we? What makes us what we are? What price are we willing to pay for this self-knowledge? These are profound questions, prompting further questions concerning the relationship among science, society, and values. Does human cloning dignify or desecrate who we are?

Human cloning powerfully reminds us of the radical nature of the connection between ontology and morality. That is, what we consider moral depends ultimately upon how we view ourselves, our purpose, and our role in a world with other selves. The questions raised by human cloning reveal all the more plainly the intimate rapport among matters of identity, meaning, and morality.

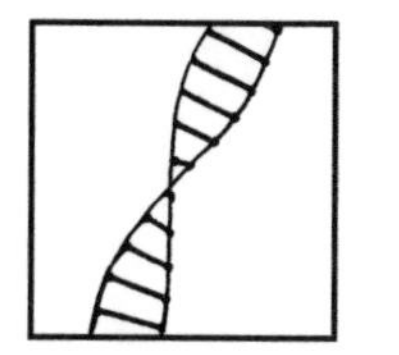

Part I

Perspectives from Science

Perspectives from Science: Introduction

IMAGES OF HUMAN cloning evoke all sorts of anxieties. Yet fears based upon misunderstanding deter us from assessing the morality of the enterprise in any sensible way; we need to know more about the science and technology behind human cloning. For instance, do the techniques of embryo splitting and somatic cell nuclear transfer pose different risks? Can what we know about natural identical twins teach us about the human clone, who is, essentially, a delayed twin? Are there sufficient differences between the natural twin and the human clone that impact upon the delayed twin's individuality and uniqueness? Moreover, human cloning compels us to address the enduring question of whether nature or nurture plays the dominant role in our development. Can we learn more about these formative influences and still avoid reductionism? And surely, selecting specific personalities for replication is difficult. But what about selecting specific phenotypes?

The underlying concern here lies in whether these questions are solely matters of science or whether they are also deeply ethical in nature, which forces us to consider an even more profound issue—the relationships between fact and value, science and morality. Are scientific facts value-free or are they value-laden, particularly with respect to human cloning? How do the scientific issues posed by human cloning affect philosophical questions of human identity and dignity? Is there a deeper meaning in what has been, throughout human history, the "natural" way of reproducing, which links sexuality and procreation? How does the new technology impact upon the relationship between parent and child and, in turn, upon society?

Jacques Cohen and Giles Tomkin point out that many of the fears associated with human cloning stem from lack of knowledge concerning its procedures and purposes. Accordingly, they aim to provide the necessary information and then go on to discuss the major areas for application of the technology. Along the way, they point out the benefits and shortcomings of cloning, claiming that it can be especially worthwhile in relieving infertility. Cloning through embryo splitting may actually pose less risk than the natural process of twinning, and the concern over possible malformations is, in their opinion, unfounded. Furthermore, it seems to be less costly than other forms of reproductive technology. In view of these benefits for infertile couples, the real question is not moral but technical: can the success rate for reproduction be improved? Cohen and Tomkin believe that it makes good sense to utilize the procedure, but they also grant that further concerns may be generated by nuclear transplantation.

Their exposé is thus scientific. For we need to have a valid empirical foundation before we can achieve any measure of moral evaluation. At the same time, the authors recognize that the ethical issues still need to be sorted out. Because most fears are unfounded and based upon misinformation, it is up to the experts to provide sensitive and sensible education to the public.

Reductionist tendencies that view development in terms of either nature or nurture too easily avoid addressing what is in fact an extremely complex process. Nevertheless, societal and scientific trends seem to focus upon nurture at one point, and later, in Hegelian fashion, to react by substituting a fascination with nature and biology, as is currently the case. In contrast to the contemporary vogue for biology and genetics, Stephen Jay Gould finds solid evidence underscoring the importance of nurture and environment in his discussions of the cloning of Dolly and of birth order's effects on human personality.

Gould also reminds us that what we now know about natural identical twins could enlighten us further about the so-called delayed twins produced by cloning. For the purer type of cloning actually occurs with natural identical twins. And the primary lesson we learn from identical twins is that nurture and environment are commanding formative factors in human development, individuality, behavior, and personhood. Despite the currently fashionable preference for the effects of nature, it is a fact that identical twins are unique, individual persons with separate identities. And, according to Gould, there are no grounds for asserting that the same will not be true for less identical, delayed twins, or clones.

Whereas Gould claims that a disposition to dichotomize can help explain the pervasive ignorance concerning human cloning, Leon Eisenberg points to the prevalence of predestination beliefs as the cause for much of the misunderstanding. He insightfully surveys early theories of human development, showing how they gave rise to the predeterminist theory. Again, the heart of the matter has less to do with ethical concerns than with correcting misleading ideas.

Eisenberg, who wrote his piece in 1976, was no doubt ahead of his time in addressing some real scientific considerations. He admits certain dangers in human cloning, if pursued on a wide scale, such as the restriction of genetic diversity, as well as problems associated with selecting the type of person to be cloned. Aside from the obvious difficulties in selecting personality traits, choosing certain phenotypes also presents some formidable challenges, such as gauging the incredibly complex interaction with an environment that never remains static. The need to avoid reifying "environment" is an issue not often raised by other commentators. Because the environment itself constantly changes, our interaction with the environment is equally dynamic. For this reason, expecting identical phenotypes is naive, even in the presence of identical genotypes.

Eisenberg thus isolates some technical problems with human cloning if its fundamental aim is to select precise phenotypes and personalities. On the other hand, if the aim in cloning is to relieve infertility, a separate set of issues arises, one major concern being the potential for physical harm via abnormal development. Nevertheless, for Eisenberg, these objections are scientific in nature and not moral.

He favors continued research while recognizing that it must proceed cautiously. Although he does not take an explicit stand either for or against human cloning, Eisenberg makes it clear that any serious discussion must not lose its focus, which is not ethical per se, but which centers on clarifying the science behind cloning.

In contrast to the positions of these scientists, Leon Kass maintains that the heart of the matter is decidedly moral and not simply technical. Kass poignantly asks, Where are the serious voices that would speak out on behalf of humanity? He claims that a good deal of current discussion lacks this serious tone, and he blames it on a postmodernist tendency to raise tolerance to the level of a virtue and to assume that any committed critique of a technological "advance" like human cloning smacks of moral absolutism. Kass challenges this contemporary concept of tolerance that encourages us to steer clear of strong moral postures.

For Kass, part of the problem lies with bioethicists who have, in effect, succumbed to the technological imperative and thus have become complicit agents of science. He contends that their analyses of ethical issues are superficial, framed by attitudes that are morally *au courant*, such as individual reproductive rights, freedom of scientific research, and matters of autonomy.

According to Kass, this context of reproductive rights and freedoms disregards more profound ontological, anthropological, and social questions. The natural process of human sexual reproduction, which links sexuality and procreation, is not simply natural and traditional; it carries deep meaning in terms of human identity and purpose. Human cloning, however, as asexual reproduction that is humanly designed, severs the significant connections evoked in the natural design. In fact, our fascination with human cloning mirrors our age in that it radically separates sexuality and procreation, and thus further fragments family ties and relations. In this way, human cloning represents the ultimate step in single parenting.

Having assigned ontological weight to the natural link between sexuality and procreation, Kass argues that human cloning jeopardizes this link in various ways. To begin with, it endangers the matter of individual identity, for possessing an *already-lived genotype* cannot help but complicate perceptions of uniqueness and familial and social identity. Human cloning also recasts reproduction as a form of *production,* and the ensuing commodification is likely to be intensified in a market-driven culture such as that of the United States. Through cloning, children stand in danger of being all the more objectified as mere instruments of others' designs and motives.

For Kass, human cloning also defiles the significance of the relationship between parent and child, bestowing a perverse new twist on the meaning of having children. Having hopes for one's child is one thing; harboring expectations biologically stamped through identical genotype is quite another. This situation is further exacerbated by our cultural preferences as to what traits are desirable. Who decides what constitutes the "perfect baby"?

As for human embryo research, Kass proposes that this endeavor can be morally sanctioned only under the strict condition that it never lead to implantation of the embryo into a uterus. His conclusion is clear: anything short of an all-out national and international ban on human cloning—a temporary moratorium or simple prohibition of federal funding for cloning research—misses the mark and amounts to an inability to take a strong and committed moral position.

The Science, Fiction, and Reality of Embryo Cloning

by Jacques Cohen and Giles Tomkin

Jacques Cohen is Professor of Embryology, Gynecology and Obstetrics and is also the Scientific Director of the Center for Reproductive Medicine and Infertility at New York Hospital, Cornell University Medical College.

Giles Tomkin is a specialist in automation and conducts IVF development and research programming at the Center for Reproductive Medicine and Infertility at New York Hospital, Cornell University Medical College.

Although many scientists view cloning as a useful procedure for scientific research into early embryo development—one that cannot currently be used to produce multiple copies of humans—the popular literature has led some individuals to view it as sinister. To address the concerns of the public, various conceptions of cloning are distinguished and their basis in fact analyzed. The possible uses, benefits, and detriments of both embryo splitting and nuclear transplantation are explained. Once the nature and purposes of cloning are understood, and the distinctive ethical dilemmas created by embryo splitting and nuclear transplantation are sorted out, these procedures should be clinically implemented to assist in vitro *fertilization treatment for those who are*

Jacques Cohen and Giles Tomkin, "The Science, Fiction, and Reality of Embryo Cloning." *Kennedy Institute of Ethics Journal*, vol. 4, no. 3 (1994): 193–203.

infertile and to further other therapeutic and investigational efforts in medicine.

MAKING IDENTICAL COPIES of molecules, cells, tissues, and even adult organisms has loosely been termed "cloning." The basic purpose of scientists in making biological copies is not generally understood. It is to avoid any variation among specimens, thereby allowing direct comparisons for scientific evaluations. Although cloning has this unambiguous rationale for investigators of biological processes, some scientific and lay persons perceive it to have unknown and sinister ends. To alleviate their fear, we wish to clarify the nature and purposes of cloning processes in general, and embryo multiplication in particular. We do so, not so much as a result of recent controversy about the prospect of human embryo cloning (Hall et al. 1993), but as a tardy response to the mistaken belief that cloning humans is currently possible. An honest discussion of the subject has probably been overdue for many years. We provide some definitions of cloning, describe the main biological processes relevant to current discussions, and consider the various uses to which cloning could be put today.[1]

Popular Conceptions of Cloning

Cloning has been defined by a number of North American reference books, works of fiction, the entertainment industry, and the scientific literature. Webster's digital dictionary explains that the word, "clone," is from the Greek *kion,* which means "twig." The term "clone" was probably first used in the botanical field to describe the process of budding. Several current uses of the term are also given, one of which is generally accepted, namely, that cloning involves creating a genetically identical individual from a single, normal, body cell. The use employed by scientists, which is less widely known, is that cloning entails asexually reproducing an identical copy of an original. Asexual reproduction is usually associated with primitive organisms such as bacteria, algae, plants, and yeast. Although it was abandoned as the only form of reproduction by such species about 500 million years ago, this method of cell division is still the only means by which the human body grows and repairs itself.

Colorful fictional depictions of human clones were developed after Aldous Huxley (1932) published his famous book on the subject, *Brave New World.* In this book, Huxley described a future society where those who continue to reproduce sexually by gestation inside the womb are designated "savages" and forced to live in special reservations; the "civilized" people are those who come to life by artificial conception and bottling as fetuses. Special laboratories provide the conditions for embryo multiplication by "budding." Birth involves "decanting" in a breeding center, and children are raised in a laboratory and conditioning center. In fact, Huxley never used the word cloning in his book.

The most controversial portrayal of cloning was given in David Rorvik's 1978 book, *In His Image: The Cloning of a Man.* Here, a millionaire called "Max" assembled a team to clone one of his body cells by inserting its genetic material into a nucleus-free egg of a donor, who then became the gestational carrier of Max's younger twin. The mixture, without acknowledgment, of facts, such as names of existing scientists, with fiction created the threatening tone of the book.[2] The book's publication coincidentally and unfairly damaged the field of assisted reproduction, which was just emerging. It has been speculated that Rorvik's book raised so much public suspicion of cloning that the American government decided in the early 1980s to halt all research on human embryos. Educated and sophisticated leaders of an advanced society were influenced by a deceptive and false tale.

The combination of *Brave New World, In His Image,* and other imaginative books, as well as science fiction fantasies in television and the cinema, generated the frightening connotations of cloning that both attract and disturb many people. These sources, which engage in a happy denial of the fact that they portray fictional and future worlds, helped to create the belief that human cloning is possible today.

If reproduction through cloning was an attractive aim, it seems unlikely that our primitive, nonhuman ancestors would have abandoned it. Modern reproductive scientists, although intrigued by the idea of creating an individual from a single body cell of an adult, have not attempted to do so because it is currently impossible.[3] Embryonic cells are initially totipotent—that is, they have the ability to develop into any kind of cell found in the body. Eventually, they lose this ability and become "differentiated" cells, such as those found in the heart or in the nervous or muscular systems. The causes of this cellular change and loss of totipotency are unknown and, therefore, cannot yet be reversed. Since cloning an individual from a single cell of an adult human being requires such reversal, it is not possible to "body-cell clone." This is the essence of the challenge that cloning presents to science. A full understanding of the process of differentiation would provide a medical magic wand, not only for cloning, but for curing or reversing malignant tumors, arterial disorders, heart disease, and ultimately the aging process.

Cloning in the Scientific Literature

The word "cloning" has appeared thousands of times in the scientific literature[4] and has been used in three broad areas. The first is genetic research, where scientists must make millions of identical copies of genes of molecular size in order to have sufficient material for testing ("molecular cloning"). The second is the production of cell-lines that have identical properties in order to study the biology of specific cells without bias from small dissimilarities between them ("cell cloning"). The third area where the word "cloning" has now appeared is that of embryo multiplication by nuclear transplantation, which is the process of introducing nuclei from the cells of early preimplantation embryos into unfertilized eggs from which the nuclei have been removed. Nuclear transplantation has been performed in early cattle embryos in order to produce nearly identical offspring with desirable traits.

The only exception to these categories within the scientific community occurs in the paper on blastomere separation of abnormal human embryos presented by Hall and co-workers to the annual meeting of the American Fertility Society in the fall of 1993 (Hall et al. 1993). There the word "clone" was used instead of "blastomere separation" to describe the artificial procedure of identical twinning.[5] In the hundreds of publications on "twinning"[6] or "splitting" embryos, this procedure has never previously been referred to as "cloning." It is usually called "blastomere separation" when the embryo is split at the four- or eight-cell stage and "embryo splitting" when the embryo is at the blastocyst stage, the last embryonic stage during preimplantation development. Scientists and society have never regarded natural twins as "clones." Consequently, the use of the word to refer to a procedure that could produce identical twins in an officially approved scientific context caused far more anxiety among members of the media and the public than the actual, fairly modest, procedures in the laboratory were intended to evoke. All that the investigators did was to separate the individual cells of abnormal embryos in order to investigate the potential of early human cells to grow independently of one another. The same work has been done many times in other species, dating back to 1932 (Nicholas 1932). This previous work is described as "blastomere separation" by some and "evaluation of totipotency" by others. The description of similar experiments performed in the human (Hardy 1990) does not include the term "cloning." Nevertheless, ever since the publicity given to the twinned embryos made in Washington, D.C., reference to embryo splitting as "cloning" is ineradicably becoming part of common usage throughout the world. Scientists therefore may now be forced to accept the term "cloning" to describe blastomere separation or embryo splitting as well as nuclear transplantation.

Kinds of Cloning

Cells from early embryos called blastomeres are each able to produce a new individual. This totipotency has allowed scientists to split animal embryos into several cells through the process of blastomere separation (Figure 1). Although this technique has been used in a number of species, it has particular advantages when applied to cattle (Willadsen 1979). In the last decade, some cattle breeding companies have used this totipotent ability to increase the success rate of their embryo transfer procedures (Gray 1991). Embryo transfer involves flushing embryos from the reproductive tract of one animal and implanting them in another. The procedure is usually successful, resulting in pregnancy rates of 60 per-

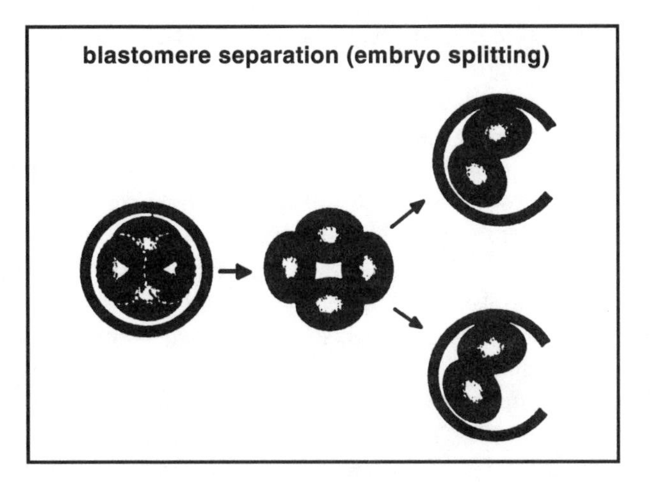

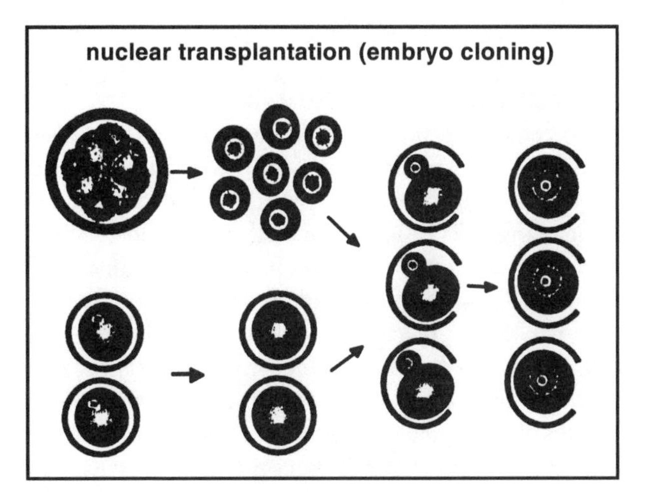

FIGURE 1

cent in the recipients. Embryo splitting can increase this percentage to 100, an improvement that is both economically and biologically advantageous, since twins can be produced from select embryos. Use of this process in cattle for more than a decade has shown no disadvantages, nor defects in the offspring.

The more technically advanced form of embryo multiplication, nuclear transplantation, was developed in the early 1980s. The genetic material from recipient eggs is removed and replaced with the nucleus of a cell from an early embryo (Willadsen 1986). This procedure, like many other manipulations on single cells or embryos, is performed using micromanipulators, instruments with microscopically fine tips that are attached to a microscope. Micromanipulators enable embryologists to perform such fine procedures as inserting the sperm cell into the egg or removing the nucleus from an embryonic cell (Cohen 1992).

In theory, large numbers of identical embryos could be produced through nuclear transplantation on multiple generations of cloned embryos, each being a descendant from a single parent eight-cell-stage embryo. The limiting factor preventing this appears to be the inability of nuclei from advanced embryos (those with hundreds of cells) to be dedifferentiated and "reprogrammed" when they are placed inside an unfertilized egg, as they have lost their totipotent capability. Nuclear transplantation has been used only in the more sophisticated animal breeding stations and is successful in a small number of laboratories. Although the procedure has been associated with higher rates of congenital malformation, evidence for this is largely anecdotal. Scientists generally agree that the birth weights of calves resulting from nuclear transplantation seem to be higher than normal. Such an increase is also being reported following the application of *in vitro* fertilization (IVF) in cattle. Further development shows no apparent abnormalities in these embryo-cloned calves with higher birth weights.

Potential Uses of Cloning in Humans

Reproductive specialists are interested in embryo duplication technology because the techniques that are currently used to treat infertile men and women are still largely unsuccessful. The national statistics indicate that only 15 percent of women undergoing IVF in 1991 gave birth to a child (Society for Assisted Reproductive Technology 1993).[7] Even this percentage is inflated when one considers that the chance of a live birth resulting from cases when a single embryo is transferred is less than 7 percent. Given these low rates of success, any advances are welcome, especially when the new technology already has been safely applied in a number of animal species. In our opinion, the morality of human embryo duplication by splitting should not be in doubt, much as no one doubts the morality of natural twinning or disputes the right of twins to existence. The real issue is whether the technique will improve the success rates of assisted reproductive technologies. Given that embryo duplication already works well in cattle, it is also very likely to work for human *in vitro* fertilization.

The primary human candidates to receive duplicated embryos would be patients using IVF procedures who have generated only a single embryo for transfer. These couples, however, currently constitute less than 10 percent of IVF patients; and even if patients who produced only two embryos were included, such couples still would constitute less than 20 percent of all IVF patients.[8] In practice, failures during the twinning procedure itself would reduce the percentage of recipients even further. Thus, the tangible benefits to this patient population would be limited. However, we believe that even if this method would improve the chances of pregnancy for patients with a single embryo by only 10 percent, the patients would undoubtedly consider it well worth undertaking. It should also be noted that the pregnancy complications associated with natural twinning are not a concern in embryo splitting. Such complications occur more often when twins share one placenta. Because the embryos produced by embryo splitting are completely separated and do not share a placenta, this problem would not arise from use of the technique.

Yet another benefit that might result from the use of embryo splitting is its potential to lower the costs associated with the use of advanced assisted reproductive technology. It is necessary to monitor patients' hormone levels and to examine their ovaries by ultrasound in the weeks leading up to the time of full oocyte maturation. Patients also must undergo a surgical outpatient procedure requiring anaesthesia in order to obtain a single batch of eggs. The use of embryo splitting would provide greater numbers of embryos for implantation from a single retrieval procedure and, therefore, would likely decrease the number of times that these procedures would have to be carried out. This, in turn, would decrease the costs of assisted reproduction. Furthermore, the laboratory work for *in vitro* fertilization includes careful monitoring of incubator conditions, a time-consuming task. Currently, the number of IVF procedures that can be performed by 100 health-care workers is about 1,000 per year. Any simplification of medical and laboratory procedures that could improve results would decrease the number of health care workers needed and ultimately also lower the cost per procedure.

Concern has been expressed that large numbers of identical children might result from embryo splitting. This is unlikely because the method has very limited numerical possibilities. After fertilization, embryonic cells only retain their totipotency through the next two, or at most, three cell divisions. According to current understanding, this permits a maximum number of four *viable* embryos to be obtained by splitting. While further splitting can be performed, it probably cannot result in viable embryos because the totipotent ability is lost, and the results will degenerate. At fertilization the cells commence to divide and do not go "backwards" when split into different embryos; that is, some sort of cellular count is kept of each cell division that occurs. Moreover, the ability to produce live offspring from the resulting embryos is also very limited. The live birth rate from the transfer of a single embryo is about 20 percent at the most advanced centers. Thus, even if 15 *healthy* embryos were successfully "twinned" from one embryo, which currently is not possible, and then singly transferred to 15 different wombs, the expected number of offspring would be only three.

There is also a concern that a greater proportion of congenital malformations would occur in children who result from the use of embryo splitting. Although this issue cannot be addressed fully until the procedures have been performed on a wide group of patients, the experience with cattle suggests that the concern is unfounded.

Assuming that the first trials of embryo splitting in patients with single embryos were successful, their success would be likely to affect all other *in vitro* fertilization procedures. If half of such procedures could be carried out using duplicated embryos, the standard number of oocytes needed to produce embryos would decrease dramatically. It would once again become feasible to use the natural menstrual cycle, rather than fertility drugs, to produce a sufficient number of oocytes for fertilization. With the prospect close at hand of simplifying procedures, increasing the success rates, and lessening the risks to patients, it seems almost immoral to some reproductive scientists not to allow embryo splitting trials for clinical use in human IVF laboratories.

Gynecologists and biologists are much more concerned about the use of nuclear transplantation to produce multiple embryos. Their concerns mostly involve unverified reports that many of the calves produced by nuclear transplantation have increased birth weights resulting in problems with their joints. As the calves mature, however, these problems seem to disappear, leaving the animals apparently normal. What these concerned scientists may have overlooked is that it is unusual to have a wide range in birth weights in cattle breeding, whereas birth weights have always varied widely in humans.[9] Therefore, an increase in birth weight in humans would not necessarily represent a problem. A second concern is that cattle breeders are not very successful when applying the nuclear transplantation technology. Although this is true, the same can also be said for practitioners of human *in vitro* fertilization.

To Clone or Not to Clone?

Ultimately, there are many good reasons to suggest that both embryo splitting and nuclear transplantation should be implemented as soon as possible in humans at the *in vitro* fertilization centers that are currently most successful. Those scientists with demonstrated expertise in performing advanced embryology should participate. Naturally, the distinctive ethical dilemmas created by nuclear transplantation will have to be sorted out before it is used to produce embryo copies.

Although the use of embryo duplication to improve the success rates of assisted reproductive technology may have immediate clinical benefit, other potential benefits of this technique may be equally significant. The technology of removing cells from embryos and transferring nuclei between cells will expand scientists understanding of the types of genes that determine the inheritance of characteristics and may also reveal how certain genetic diseases are activated. Other benefits seem sure to result from studying how and why embryonic cells differentiate into the specific cells of different organs. This complex process can be imitated by producing stem-cell lines from embryonic cells. The study of these stem cells is important for expanding scientists' understanding of many diseases.

We believe that irrational fears and fictional scenarios currently play a dominant role in determining ethical attitudes toward embryo splitting. The word "clone" has a negative and misleading significance for most people. Scientists and medical health-care workers, especially those who specialize in embryology and genetics, have erred by not informing the public of the real nature of cloning and its uses in assisted reproductive technology. Their own familiarity with concepts such as nuclear manipulation and genetic transfer and the great needs of the individual infertility patient has led them, all too frequently, to forget that outsiders, both lay and professional, may be disturbed by discussion of these realities. It is often easy to overlook the confusion between concepts that affects even professional practitioners in the absence of clear definitions of terms. When such confusion can occur among professionals in the field, it obviously can occur more extensively among groups without their background knowledge. It is hoped that the clarification of terms and ideas that we have presented will serve to enhance public understanding of the nature and purpose of cloning.

Notes

1. The views contained in this article are our own and may represent a minority opinion in certain areas. Where applicable, we refer to views held by others in the field.
2. The publishers acknowledged in their foreword that they were not sure if the story were true since the persons involved wanted to remain anonymous.
3. Nuclei from adult mammalian cells, when transferred to fresh enucleated eggs from the same species cannot commence cell division and chromosome duplication and therefore do not develop. The furthest development among lower species has been accomplished in the tadpole experiments of Dr. John Gurdon (1986) from Cambridge University in the United Kingdom.
4. This estimate was derived from Mini-Medline searches and publication archives.
5. Natural identical twins are also derived from a single embryo. The process should not be confused with fraternal twinning in which multiple eggs (each genetically different) are released from the ovary and fertilized by different sperm. Fraternal twins are not genetically identical.
6. It is not certain that Hall's experiment produced identical twins. The embryonic cells that developed following separation of the abnormally fertilized egg were never tested for genetic homogeneity, which is crucial to the establishment of proof that the half-embryos were indeed identical. The abnormal embryo used in these experiments was fertilized by two sperm, which is known to cause mosaicism in animal as well as human embryos. The individual cells of a mosaic embryo are by definition all genetically different from each other because of errors of nuclear divisions (Pieters 1992; Long 1993).
7. The Society for Assisted Reproductive Technology, which is part of the American Fertility Society, surveys more than 200 member programs each year.
8. This estimate was derived from analyses performed on embryos from 1200 patients using a relational database (EggCyte) at the program of Cornell University Medical Center.
9. This opinion was presented by Dr. Steen Willadsen at the workshop on embryo cloning held by the National Advisory Board on Ethics in Reproduction, Washington, D.C., 15 February 1994.

References

Cohen, Jacques; Malter, Henry E; Talansky, Beth E.; and Grifo, Jaime. 1992. *Micromanipulation of Human Gametes and Embryos.* New York: Raven Press.

Gray, K. R.; Bondioli, K. R.; Betts, C. L. 1991. The Commercial Application of Embryo Splitting in Beef Cattle. *Theriogenology* 35: 37–44.

Gurdon, J. B. 1986. Nuclear Transplantation in Eggs and Oocytes. *Journal of Cell Science* 4 (supplement): 287–318.

Hall, J. L.; Engel, D.; Gindoff, P. R.; et al. 1993. Experimental Cloning of Human Polyploid Embryos Using an Artificial Zona Pellucida. The American Fertility Society conjointly with the Canadian Fertility and Andrology Society, Program Supplement, 1993 Abstracts of the Scientific Oral and Poster Sessions, Abstract 0-001, S1.

Hardy, Kate; Martin, Karen L.; Leese, Henry J.; et al. 1990. Human Preimplantation Development In Vitro Is Not Adversely Affected by Biopsy at the 8-cell Stage. *Human Reproduction* 5: 708–14.

Huxley, Aldous. 1932. *Brave New World.* London: Chatto and Windus. Long, C. R.; Pinto-Correia, C.; Duby, R. T.; et al. 1993. Chromatin and Micro-tubule Morphology During the First Cell Cycle in Bovine Zygotes. *Molecular Reproduction and Development* 36: 23–32.

Nicholas, J. S. 1932. Development of Transplanted Rat Eggs. *Proceedings of the Society of Experimental Biology and Medicine* 30:1111–27.

Pieters, M. H. E. C.; Dumoulin, J. C. M.; Ignoul-Vanvuchelen, R. C. M.; et al. 1992. Triploidy after In Vitro Fertilization: Cytogenetic Analysis of Human Zygotes and Embryos. *Journal of Assisted Reproduction and Genetics* 9:68–76.

Rorvik, David. 1978. *In His Image: The Cloning of a Man.* Philadelphia: Lippincott.

Society for Assisted Reproductive Technology, The American Fertility Society. 1993. Assisted Reproductive Technology in the United States and Canada: 1991 Results from the Society for Assisted Reproductive Technology, American Fertility Society Registry. *Fertility and Sterility 59: 956–62.*

Willadsen, Steen M. 1979. A Method of Culture of Micromanipulated Sheep Embryos and Its Use to Produce Monozygotic Twins. *Nature* 277: 298–300.

———. 1986. Nuclear Transplantation in Sheep Embryos. *Nature* 320: 63–65.

The Outcome as Cause: Predestination and Human Cloning*

by Leon Eisenberg

Leon Eisenberg is the Maude and Lillian Presley Professor of Psychiatry at Harvard Medical School.

The demonstration in the laboratory that vertebrates can be cloned has been taken to imply that it is but a matter of time and technique until such is possible in man. Were human cloning to be accomplished, it has been assumed that the result would be psychological as well as physical identity between the originator of the clone and his or her cloned offspring. This unwarranted assumption reveals the persistence of the concept of predestination as a causal force in development, despite its incompatibility with biological evidence. The basic problems with the fiction of human cloning are not ethical or political but scientific. Pseudoscience serves to distract attention from fundamental social and moral questions.

"The Outcome as Cause: Predestination and Human Cloning," by Leon Eisenberg, *The Journal of Medicine and Philosophy*, vol. 1, no. 4 (1976): 318–31.

*An earlier version of this paper, entitled "The Psychopathology of Clonal Man," was presented at the National Symposium on Genetics and the Law held in Boston, May 18–20, 1975, and was published in *Genetics and the Law*, edited by A. Milunsky and G. J. Annas (New York: Plenum Press, 1975).

Experiments in Cloning

A CLONE IS the aggregate of the asexually produced progeny of one individual. Reproduction by cloning in horticulture employs cuttings of a single plant to propagate desired botanical characteristics indefinitely. In microbiology, a colony of bacteria constitutes a clone when its members are the descendants of a single bacterium which has undergone repeated asexual fission; the myriad bacteria in a clone have each precisely the same genetic complement as that of the progenitor cell and are indistinguishable one from another.

Recently, it has proved possible to "clone" copies of amphibia *(Rana pipiens* and *Xenopus laevis)* by the technique of transferring nuclei from an embryo or a tadpole (Briggs and King 1952; Gurdon, Elsdale, and Fischberg 1958; Gurdon 1962, 1968). The first step in the method is to remove the nucleus of a frog egg by microdissection or to destroy it by ultraviolet radiation. A differentiated (intestinal epithelial) cell from a second frog is sucked into a micropipette in such a manner as to destroy the cell wall and remove much of the cytoplasm. The nucleus (and a small amount of residual cytoplasm) from the differentiated cell is then injected into the recipient enucleated egg. If conditions are right, the diploid nucleus from the second frog initiates embryogenesis in the egg from the first. In about 1–2 percent of trials, this process continues through all the intermediate developmental stages to the production of the mature animal. Since this new frog has chromosomes identical with those of the nuclear donor, it is a member of a clone which can be replicated indefinitely (at least in theory) by the technique of successive transfer of cell nuclei into recipient eggs.

These extraordinary experiments were undertaken to explore the factors that control embryonic differentiation. They demonstrate unequivocally that at least some of the nuclei in fully differentiated amphibian cells do not "shed" or "lose" genes but contain the full complement of potentially active genetic material that is present in the zygote; what distinguishes differentiated cells in which sets of genes are turned "off" or "on."

In addition to their biological significance for embryogenesis, these experiments have suggested that similar techniques might make it possible to "clone" human beings (by transferring a human ovum to a test tube, removing its nucleus, replacing it with a somatic cell nucleus from a "suitable" donor, allowing it to differentiate to the blastula stage and then implanting it in a "host" uterus). Such an individual, if it attained maturity, would be an identical genetic twin of the adult nuclear donor. This hypothetical outcome, although remote, has given rise to speculation about the psychological, ethical, and social consequences of producing clones of identical human beings (Ramsey 1970). The futuristic scenarios evoked by the prospect of human cloning contain implicit assumptions about the causes of development. Examination of these underlying premises highlights themes that can be traced back to Greek antiquity. Nonetheless, they continue to have salience for understanding contemporary debates about the sources of individual and group differences in such characteristics as "intelligence" and "aggression" (Eisenberg 1972; Ludmerer 1972; Block and Dworkin 1976).

THEORIES OF DEVELOPMENT

The enigmas of development have concerned philosophers and naturalists since man first began to wonder how plants and animals emerged from the products of fertilization (Needham 1959). There is no resemblance between the physical appearance of the seed and the form of the adult organism. Yet the plant or animal to which it gives rise is an approximate replica of its progenitors. The earliest Greek explanatory concept was the doctrine of preformationism—that is, the unfolding of preexisting structures which, although they cannot be seen by the naked eye, must be supposed to have been present in the seed. This ancient speculation, found in the hippocratic corpus, was given poetic expression by the Roman Seneca: "In the seed are enclosed all the parts of the body of the man that shall be formed. The infant that is borne in his mother's womb has the roots of the beard and hair that he shall weare one day. In this little masse likewise are all of the lineaments of the body and all that which posterity shall discover in him" (*Questiones naturales,* trans. T. Lodge, 1610; quoted in Needham [1959], p. 66).

Preformationism was so powerful an ideology that, when the microscope was invented two millennia later, the first microscopists to examine a sperm were able to persuade themselves that they could see in its head a homunculus with all the features of a tiny but complete man. The introduction of better technology and the beginnings of embryology as an experimental science made the doctrine progressively more difficult to sustain in its original form. With better microscopic resolution, the expected structures could not be seen; experimental manipulation of embryos produced abnormal "monsters" which could not have been already present in the seed.

In the seventeenth and eighteenth centuries, preformationism evolved into predeterminism, a view that infused the Greek belief with the Christian doctrine of predestination. This position was epitomized by Swammerdam (quoted in Needham 1959, p. 170): "In nature, there is no generation but only propagation, the growth of parts. Thus original sin is explained, for all men were contained in the organs of Adam and Eve. When their stock is finished, the human race will cease to be."

Although few contemporary biologists continue to speak of original sin, many adhere to predeterminism, now restated in the guise of the early (and no longer tenable) versions of genetic theory. Current polemics on the inheritance of intelligence philosophically reiterate the ancient premises of predeterminism and predestination (Block and Dworkin 1976).

The alternative view, that of epigenesis, was formulated by Aristotle in opposition to preformationism. Having opened eggs at various stages in development, he observed that the individual organs did not all appear at the same time, as preformationist theory demanded. He did not accept the argument that differences in the size of the organs could account for the failure to see them all at the same time. Others as well as he had noted that the heart is visible before the lungs, even though the lungs are ultimately much larger. Unlike his predecessors, Aristotle began from the observable data. He concluded that new parts were formed in succession and did not merely unfold from precursors already present:

> It is possible, then, that A should move B and B should move C, that, in fact, the case should be the same as with the automatic machines shown as curiosities. For the parts of such machines while at rest have a sort of potentiality of motion in them, and when any external force puts the first of them into motion, immediately the next is moved in actuality. As, then, in these automatic machines the external forces moves the parts in a certain sense (not by touching any part at the moment but by having touched one previously), in like manner also that from which the semen comes, or in other words that which made the semen, sets up the movement in the embryo and makes the parts of it by having touched first something though not continuing to touch it. In a way it is the innate motion that does this, as the act of building builds a house. Plainly, then, while there is something which makes the parts, this does not exist as a definite object, nor does it exist in the semen at the first as a complete part. [Quoted in Needham 1959, pp. 47–48]

This is the first statement of the theory of epigenesis; successive differentiations in the course of development give rise to new properties and new structures.

It is, of course, still true that the genetic code in the zygote determines the range of possible outcomes; it is only a fertilized human ovum that can give rise to a human infant. Yet the genes active in the zygote serve only to initiate a sequence, the outcome of which is dependent on the moment-to-moment interactions between the products. of successive stages in development. For example, the potentiality for differentiating into pancreas is limited to cells in a particular zone of the embryo. But they will produce prozymogen, the histologic marker for pancreas, only if they are in contact with neighboring mesenchymal cells; dissected away from mesenchyme, their evolution is arrested despite their genetic "potential" (Grobstein 1964). And at the same time, the entire process is dependent on the adequacy of the uterine environment, defects in which lead to anomalous development and miscarriage.

The Biology of Reproduction

The methodologic barriers to successful human cloning are formidable. Even in amphibia, development proceeds to maturity only one time in 100 (Gurdon 1968). Its accomplishment in mammals has yet to be demonstrated. Nonetheless, even if the necessary virtuosity lies in the more distant future than science fiction enthusiasts suggest, for the sake of discussion we can state that a solution exists in principle (much like the mathematician who, observing his house on fire and noting water in a nearby pool, decided that he could return to his studies, now that the fire had been extinguished in principle!) and attempt to envisage the possible outcomes.

Wide-scale cloning would invite *biological* disaster; for it would lead to a marked restriction in the diversity of the human gene pool. Such a limitation would endanger the ability of our species to survive major environmental changes. Genetic homogeneity is compatible only with adaptation to a very narrow ecological niche. Once that niche is perturbed (through an invasion by a new predator, by a change in temperature or water supply, etc.), extinction follows. The risks constitute a clear and present danger. The "green revolution" in agriculture has led to the selective cultivation of grain seeds chosen for high yield under modern conditions of fertilization and pest control. Worldwide food production, as a result, is now highly vulnerable to the appearance of new blights because of our reliance on a narrow range of genotypes (Harlan 1975). Recognition of this threat has led to a call for the creation of seed banks containing representatives of "wild" species as protection against catastrophe from new blights or changed climatic conditions to which current high-yield grains prove particularly vulnerable (National Academy of Sciences 1972). Precisely the same threat would hold for man, were we to replace sexual reproduction by cloning.

The extraordinary biological investment in sexual reproduction (as compared with asexual replication) provides a measure of its importance to species evolution (Wilson 1975). Courtship is expensive in its energy requirements; reproductive organs are elaborate; there are extensive differences between male and female in secondary sexual characteristics. These have evolved (Darwin 1871) as the result of sexual selection (by successive choice of mating partners and competition between males for access to the female). The benefit of sexual reproduction is the enhancement of diversity (by crossover between homologous chromosomes during meiosis and by combining the haploid gametes of a male and a female parent). The new genetic combinations so produced enable the species to respond as a population to changing environmental conditions by selective survival of adaptable genotypes. The deliberate imposition of a restriction on human genetic diversity would violate the very biological principles basic to the evolution of our species.

Donor Selection

There is a further dilemma in the task of selecting the "ideal types" to be cloned. That choice could only be made on the basis of phenotypic characteristics which had arisen under the environmental conditions present during the several decades in which these individuals had come to maturity. Let us set aside the problem of value choice: do we emphasize intellectual dexterity, competitiveness or cooperativeness, achievement orientation, empathy, artistic ability, or certain permutations and combinations thereof? For the sake of argument, assume agreement on the traits to be valued, however unrealistic that assumption. By definition, the "genetic potential" for those characteristics must have existed in the individuals who now exhibit them. But the translation of that hypothetical genotype into this visible phenotype occurs in a complex interaction with the environment present during development. Even if we agree on the bearers of the genotype we wish to preserve, we face a formidable barrier: we know so little of the environmental features necessary for the flowering of that genotype that we cannot specify in detail the environment we would have to provide, both pre- and postnatally, to assure a phenotypic outcome identical with the complex of traits we aim to perpetuate.

Again, it is possible to contend that, even if we do not know that now, we will in the future; let us make a further dubious assumption and suppose a day has arrived when we can specify the immediate environment necessary for the phenotypic flowering of the chosen phenotype. Nonetheless, the phenotype so admirably suited to the world of the past in which it matured may not be adaptive to a world a generation hence. That is, the traits which lead an individual to be creative or to exhibit leadership at one moment in history may not be appropriate at another.

Not only is the environment not static; it is altered by our own extraordinary impact on our ecology. The proliferation of our species changes disease patterns (Black 1975); our methods of disease control, by altering population ratios, can affect the physical environment itself (Omerud 1976); social revolutions demand new types of leaders. Cloning would condemn us ever to plan the future on the basis of the past (since the successful phenotype cannot be identified sooner than adulthood). Our inability to anticipate the future makes it impossible to gauge the fitness of the chosen genotype for the generation hence when it will have ripened into its phenotype. Advocates of human cloning, no less than the critics who start from the same erroneous biological premises in order to argue against it on ethical grounds, fail to understand development as an epigenetic process.

Genotype and Phenotype

For the student of biology, cloning is a powerful and instructive method with great potential for deepening our understanding of the mechanisms of differentiation during development. The norm of reaction (range) of the genotype is a measure of its multivariate characteristics in dynamic interaction with the environment. That norm can only be estimated from the varied manifestations of the phenotype over as wide a range of environments as permit its survival. The more varied those environments, the greater the diversity observed in the phenotypic manifestations of the one genotype. Human populations possess an extraordinary range of latent variability. Dissimilar genotypes can produce remarkably similar phenotypes under the wide range of conditions that characterize the environments of the inhabitable portions of the globe. The differences resulting from genotypic variability manifest themselves most clearly under extreme conditions when severe stresses overwhelm the homeostatic mechanisms which ordinarily buffer the system against small perturbations.

To repeat, for identity in phenotype, we require identity in genotype, which cloning can assure, *and* identity in environmental interactions. At the most trivial level, we can anticipate less similarity even in physical appearance between cell donor and cloned recipient than that which is observed between one-egg twins. Inequalities in placentation and fetal-maternal circulation can vary significantly, even for uniovular twins housed in one uterus (Benirschke 1972). Developmental circumstances will be more variable between donor and "clonee," who will have been born from different uteri.

Let us force the argument one step further. Let us assume that the conditions for the cloned infant will have been identical with those of his or her progenitor, so that at birth the infant will be a replica of its "father" or "mother" at birth. Under such circumstances (and within the limits of precision of genetic specification), the immediate pattern of central-nervous-system connections and their response dispositions to stimulation will be the same as those of the progenitor at birth.

The Molding of Brain Structure

However, even under these circumstances, the future is not predestined. The human species is notable for the proportion of brain development that occurs postnatally. Other primate brains increase in weight from birth to maturity by a factor of 2.2–2.5 (orangutan), but the brain of man increases by 3.5–4.0. There is a four-fold increase in the neocortex, a marked elaboration of the receiving areas for the teloreceptors, a disproportionate expansion of the motor area for the hand in relationship to the representation for other parts, a representation of tongue and larynx many times greater, and a great increase in the "association" areas. The elaboration of pathways and interconnections is highly dependent on the quantity, quality, and timing of intellectual and emotional stimulation (Eisenberg 1975*a*). The very structure of the brain as well as the function of the mind emerge from the interaction between maturation and experience (Aronson et al. 1972). In the human case, this is hypothesis rather than established fact.

The density of cell and fiber structure in the human brain eludes quantitative analysis by current neurobiological methods of anatomic differences resulting from experience.

But the hypothesis that experience molds structure is so strongly supported by experiments with animals that its extrapolation to man is justified. In the kitten, at birth and prior to eye opening, the visual system of the newborn already displays precise organization and functional competence (Hubel and Wiesel 1963; Wiesel and Hubel 1963, 1965). Some 80 percent of the cells in the striate cortex can be driven by either eye. If one eye is deprived of light (or even of patterned vision) for some weeks while the other eye is free to see, the restricted eye loses its capacity to drive most of the previously binocular cells, and there is significant cell shrinkage in the intermediate way stations. Yet if both eyes are restricted, unilateral "capture" does not occur, and the striate cells respond to each eye when the occlusion is removed. Thus at birth the kitten possess a visual "wiring diagram" fashioned by the genetic code under average conditions of intrauterine development. That preliminary wiring diagram is dependent upon external stimulation for the maintenance of its salient features (e.g., binocularity); it is fashioned into its ultimate adult circuitry by the patterns of stimulation falling upon it.

Similar phenomena are evident in man. If strabismus is not corrected before the fifth year in children, the capacity for fine visual discrimination is lost in one eye. An even more remarkable example is provided by the case of congenital astigmatism. Asymmetries in the lens make it impossible for light to be sharply focused on the retina in the horizontal or the vertical meridian. With special optical devices in the laboratory, it is possible to bypass the lenticular problem and focus light precisely on the retina. Yet in astigmatic adults, the retina no longer exhibits normal visual acuity in the meridian of prior error, even when the optical stimulus is precisely aligned (Freedman, Mitchell, and Milodot 1972). Abnormal input during the early years of development has permanently modified the receptive fields and synaptic properties of the visual system. If sensory deprivation impoverishes brain structure (Riesen 1966), complex physical and social environments enhance it (Globus et al. 1973; Greenough 1975). The richer the environment, the greater the complexity of dendritic branching and synaptic junctions and the more efficient the learning when the animal is exposed to novel situations. These studies serve to emphasize not only the growth-dependent processes of maturation but also the role of learning and experience in social ontogeny (Schneirla and Rosenblatt 1961).

There is yet another level of complexity in the analysis of personality development. The human traits of interest to us are polygenic rather than monogenic; similar outcomes can result from the product of the interaction between different genomes and different social environments. To produce another Mozart, we would need not only Wolfgang's genome but mother Mozart's uterus, father Mozart's music lessons, their friends and his, the state of music in eighteenth-century Austria, Hayden's patronage, and on and on, in ever-widening circles. Without his set of genes, the rest would not suffice; there has been, after all, only one Wolfgang Amadeus Mozart. But we have no right to the converse assumption: that his genome, cultivated in another world at another time, would result in an equally creative musical genius. If a particular strain of wheat yields different harvests under different conditions of climate, soil, and cultivation, how can we assume that so much more complex a genome as that of a human being would yield its desired crop of operas, symphonies, and chamber music under different circumstances of nurture?

I have thus far identified three objections to cloning: its devastating impact on the diversity of the human gene pool, were it to be widely adopted; its inevitable selection for traits successful in the past and not necessarily adaptive to an unpredictable future; and the vulnerability of the phenotype it might produce, given the uncontrollable vicissitudes of the environment it will encounter. These seem to me to be the fundamental objections to human cloning, it if were to be pursued as a strategy for "positive" eugenics by "genetic engineering."

Clinical Applications

If, on the other hand, we look at cloning in the context of individual patient care, then the relevant issues are substantially different. Consider the partners in a barren marriage. If the husband is sterile, the couple might wish to opt for reproduction via the transfer of a cell nucleus from the husband to an enucleated egg from the wife, with the resulting embryo to be implanted in the wife's uterus. They might prefer this choice to artificial insemination. It would result, of course, in a male infant genetically identical with the father. In like fashion, a somatic nucleus from the mother could be implanted in one of her own ova if they wanted to have a daughter. If the wife is anovulatory, a "donated" ovum could serve as the vehicle for cloning. The egg with a transferred nucleus cultivated in vitro prior to implantation would, in other respects, be no different from one produced by joining father's (or donor's) sperm and mother's (or donor's) ovum in vitro to overcome infertility. All such procedures carry hazards of maldevelopment. The risks from induction by nuclear transfer would have to be weighed against the available alternatives, each of which carries risk (for example, clomiphene may induce multiple births). A cloned child and its parents might be in for a difficult time if they harbored unrealistic expectations of genetic determinism, but such is already the case with biological parents preoccupied with "bad seed" in the family. If the donor of the nucleus were not a family member, the relationship between the rearing parent and the child would not differ genetically from that in extrafamilial adoption. One can well question the propriety of the whole enterprise, in view of the availability of adoption as a socially preferable enterprise. While it may be true that Darwinian evolution has selected for reproductive behaviors that enhance the likelihood of gene transmission (Trivers 1972), the elaboration of human consciousness and culture enables parental and filial commitment to be lasting in the absence of genetic similarity.

The Politics of Science

Proposals for human cloning as a method for "improving" the species err because they are based on biological nonsense; to elevate the question into an ethical issue is sheer casuistry. The problem lies not in the "nature" of clonal man but in the metaphysical cloud that surrounds this hypothetical creature. Pseudobiology provides a platform for trivializing ethics and distracts attention from real and present injustice.

It has become a fashionable academic exercise to speculate that despotic governments might employ the "technology" of cloning by selecting appropriate donors to create docile populations. That goal need not await a new methodology; leaders so motivated could employ the techniques of animal husbandry for selective breeding. Both methods have little to commend them to a dictator. They are absurdly complicated, take far too long, and are uncertain in their outcome. False beliefs, disseminated and enforced by the state, quite suffice. The power of mythopoesis was understood with remarkable clarity by the Athenian philosophers.

In Plato's *Republic,* Socrates (in a passage prescient of Watergate) comments that, "if anyone at all is to have the privilege of lying, the rulers of the State should be the persons; and they, in their dealings with the enemy or with their own citizens, may be allowed to lie for the public good. But nobody else should meddle with anything of the kind . . . " (p. 89).

He then goes on to describe to Glaucon a "needful falsehood," "an audacious fiction" which will assure the loyalty of citizens to the state:

> Citizens, we shall say to them in our tale, you are brothers, yet God has framed you differently. Some of you have the power to command, and in the composition of these he has mingled gold, wherefor also they have the greatest honor; others he has made of silver, to be auxiliaries; others again who are to be husbandmen and craftsmen he has composed of brass and iron; and the species will generally be preserved in the children. . . . God proclaims as a first principle to the rulers, and above all else, that there is nothing which they should so anxiously guard, or of which they are to be such good guardians, as of the purity of the race. . . . For an oracle says that when a man of brass or iron guards the State, it will be destroyed. . . . [P. 129]

Later he comments, "Seeing then that there are three distinct classes, any meddling of one with another, or the change of one into another, is the greatest harm to the State, and may be most justly termed evil doing . . . " (p. 155). Will the tale be believed? The present generation may not be persuaded, "But their sons may . . . and their sons' sons, and posterity after them" (p. 130).

Will forbidding scientists to experiment with cloning restrain the state from the unjust exercise of its powers? Is it not more likely that the search for truth will expose the myths of gold, silver, brass and iron, the myths of genetic predestination, which justify inequality? Rulers and slaves existed long before ancient Athens. Then as now, the police powers of the state quite sufficed to stratify society into castes and classes; audacious scientific fictions provided the useful lies to rationalize what was as what should have been. The division of mankind into master races and inferior subspecies—and the assertion that both breed true—is as old as human history and as current as Shockley and Jensen. Jesuitical disquisitions about the status of the clonal homunculus provide safe and genteel occupation for those better employed in the struggle against the violations of the rights of real women and men.

To a generation reared in the shadow of nuclear catastrophe and confronting the despoliation of the environment, it is evident that the fruits of applied science are sometimes bitter and the aftertaste delayed. Man's ambivalence about knowing is an ancient theme. Greek mythology abounds in minatory legends of man's punishment for daring to steal fire from the gods, for fashioning wings to fly, for venturing into the realms of the dead. Western religious tradition tells us man was cast out of the Garden of Eden for having eaten from the Tree of Knowledge. Faust must sell his soul to the devil in his quest for wisdom and immortality.

Legends and dogma delayed but could not halt the power of the new science. Grudgingly, the goods coveted by all were reinterpreted as signs of God's bounty, although with constant caveats that there are limits we dare not transcend. As our age discovers that we sometimes transmute gold into dross in the technological imperative to do what we have the means to do without weighing the consequences, old doctrines emerge in a new guise. Because science is incomplete, reason imperfect, and both suborned by the appetites of the state, some would abandon science and reason in favor of mysticism, hermetics, and transcendental rapture. Particularly is this so when the new biology confronts the greatest wonder of all: the nature of life itself.

Of course, we should consider what we are about to do before we attempt it. Virtuosity is not its own justification; human values take precedence. But I reject categorically the proposal that we turn back at the edge of greater understanding of the biology of life, an understanding which can increase our dominion over ourselves. Scientific studies of cloning in the laboratory offer a powerful tool for exploring the mechanisms of embryonic differentiation. They hold promise of stripping away the mysteries of sequential translation of the genetic code in the process of development. In what sense is in vitro cultivation of the beginnings of wanted life any more "unnatural" than contraception and abortion to prevent unwanted progeny? Some condemn them all. I confess myself utterly unable to understand the "ethics" that regard contraception and abortion as abominations without blinking at the sight of women crippled by illegal abortions, of infants neglected and abandoned, and children starving (Eisenberg 1975*b*). The "right-to-lifers" who applauded the conviction of Dr. Edelin for aborting a twenty-two-week-old fetus stand ready to stone the same black child bused into their neighborhoods!

It is no less natural that we now possess the means to prevent famine, if we had the moral commitment to do it, than that once we were its helpless victims. It is no more natural that we beggar ourselves by overpopulation than that we learn to plan parenthood. Because God or Nature joined procreation with the sexual act, are we to be forbidden to rend them asunder by artificial insemination or fertilization in vitro? I find the logic as persuasive as the proposition that the use of antibiotics violates our moral covenant because God or Nature intended the pneumococcus no less than man.

It is not knowledge but ignorance that assures misery. It is not science but its employment for inhuman purposes that threatens our survival. The fundamental ethical questions of science are political questions: who shall control its products? For what purposes shall they be employed?

The ethical blight in treatises on clonal man lies in the preoccupation with biological myths about genetic predestination in the place of moral concern for the social realities of caste and class.

References

Aronson, L. R.; Tobach, E.; Lehrman, D. S.; and Rosenblatt, J. S., eds. *Selected Writings of T. C. Schneirla.* San Francisco: W. H. Freeman Co., 1972.

Benirschke, K. "Multiple Births." In *Pediatrics,* edited by H. L. Barnett. New York: Appleton-Century-Crofts, Inc., 1972.

Black, F. L. "Infectious Diseases in Primitive Societies." *Science* (February 14, 1975), pp. 515–18.

Block, N. J., and Dworkin, G. *The I.Q. Controversy.* New York: Pantheon Books, 1976.

Briggs, R., and King, T. J. "Transplantation of Living Nuclei from Blastula Cells into Enucleated Frogs' Eggs." *Proceedings of the National Academy of Sciences of the United Slates of America* (May 15, 1952), pp. 455–63.

Darwin, C. *The Descent of Man and Selection in Relation to Sex,* 2 vols. New York: Appleton, 1871.

Eisenberg, L. "The *Human* Nature of Human Nature." *Science* (April 14, 1972), pp. 123–28.

Eisenberg, L. "Primary Prevention and Early Detection in Mental Illness." *Bulletin of the New York Academy of Medicine* 51 (January 1975): 118–29. (*a*)

Eisenberg, L. "Acting amidst Ambiguity: The Ethics of Intervention." *Journal of Child Psychology and Psychiatry and Allied Disciplines* 16 (April 1975): 93–104. (*b*)

Freedman, R. D.; Mitchell, D. E.; and Millodot, M. "A Neural Effect of Partial Visual Deprivation in Humans." *Science* (March 24, 1972), pp. 138–86.

Globus, A.; Rosenzweig, M. R.; Bennett, E. L.; and Diamond, M. C. "Effects of Differential Experience on Dendritic Spine Counts in Rat Cerebral Cortex." *Journal of Comparative and Physiological Psychology* 82 (February 1973): 175–81.

Greenough, W. T. "Experiential Modification of the Developing Brain." *American Scientist* 63 (January-February 1975): 37–46.

Grobstein, C. "Cytodifferentiation and Its Controls." *Science* (February 14, 1964), pp. 643–50.

Gurdon, J. B. "The Developmental Capacity of Nuclei Taken from Intestinal Epithelium Cells of Feeding Tadpoles." *Journal of Embryology and Experimental Morphology* 10 (December 1962): 622–40.

Gurdon, J. B. "Transplanted Nuclei and Cell Differentiation." *Scientific American* 29 (December 1968): 24–35.

Gurdon, J. B.; Elsdale, T. R.; and Fischberg, M. "Sexually Mature Individuals of *Xenopus laevis* from the Transplantation of Single Somatic Nuclei." *Nature* (July 5, 1958), pp. 64–65.

Harlan, J. R. "Our Vanishing Genetic Resources." *Science* (May 9, 1975), pp. 618–21.

Hubel, D. H., and Wiesel, T. N. "Receptive Fields of Cells in Striate Cortex of Very Young, Visually Inexperienced Kittens." *Journal of Neurophysiology* 26 (November 1963): 994–1002.

Ludmerer, K. M. *Genetics and American Society.* Baltimore: Johns Hopkins Press, 1972.

National Academy of Sciences. *Genetic Vulnerability of Major Crops.* Washington, D.C.: National Academy of Sciences, 1972.

Needham, J. *A History of Embryology.* New York: Abelard/Schuman, Ltd., 1959.

Omerud, W. E. "Ecological Effect of Control of African Trypanosomiasis." *Science* (February 27, 1976), pp. 815–21.

Plato. *The Republic.* In *The Works of Plato,* translated by B. Jowett. New York: Tudor Publishing Co., n.d.

Ramsey, P. *Fabricated Man: The Ethics of Genetic Control.* New Haven, Conn.: Yale University Press, 1970.

Riesen, A. H. "Sensory Deprivation." In *Progress in Physiological Psychology. I,* edited by E. Stellar and J. M. Sprague. New York: Academic Press, 1966.

Schneirla, T. C., and Rosenblatt, J. S. "Behavioral Organization and Genesis of the Social Bond in Insects and Mammals." *American Journal of Orthopsychiatry* 31 (April 1961): 223–53.

Trivers, R. L. "Parental Investment and Sexual Selection." In *Sexual Selection and the Descent of Man, 1871–1971,* edited by B. Campbell. Chicago: Aldine-Atherton Publishing Co., 1972.

Wiesel, T. N., and Hubel, D. H. "Single Cell Responses in Striate Cortex of Kittens Deprived of Vision in One Eye." *Journal of Neurophysiology* 26 (November 1963): 1003–17.

Wiesel, T. N., and Hubel, D. H. "Comparison of the Effects of Unilateral and Bilateral Eye Closure on Cortical Unit Responses in Kittens." *Journal of Neurophysiology* 28 (November 1965): 1029–40.

Wilson, E. O. *Sociobiology: The New Synthesis.* Cambridge, Mass.: Harvard University Press, 1975.

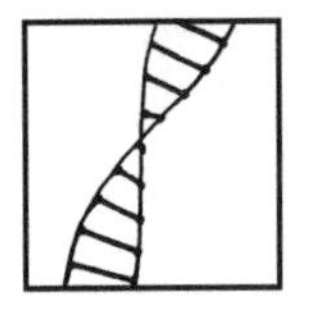

Dolly's Fashion and Louis's Passion: Ruminations on the Downfall of a King and the Cloning of a Sheep

by Stephen Jay Gould

Stephen Jay Gould is Professor of Paleontology and Zoology at both Harvard and New York Universities. He is also former president of the American Academy for the Advancement of Science.

NOTHING CAN BE more fleeting or capricious than fashion. What, then, can a scientist, committed to objective description and analysis, do with such a haphazardly moving target? In a classic approach, analogous to standard advice for preventing the spread of an evil agent ("kill it before it multiplies"), a scientist might say, "quantify before it disappears."

Francis Galton, Charles Darwin's charmingly eccentric and brilliant cousin, and a founder of the science of statistics, surely took this prescription to heart. He once decided to measure the geographic patterning of female beauty. He attached a piece of paper to a small wooden cross that he could carry, unobserved, in his pocket. He held the cross at one end in the palm of his hand and, with a needle secured between thumb and forefinger, made pinpricks on the three remaining projections (the two ends of the crossbar and the top).

He would rank every young woman he passed on the street into one of three categories—as beautiful, average, or substandard (by his admittedly subjective

---

Reprinted with permission from *Natural History* (June 1997). Copyright by the American Museum of Natural History (1997).

preferences)—and he would then place a pinprick for each woman into the designated domain of his cross. After a hard day's work, he tabulated the relative percentages by counting pinpricks. He concluded, to the dismay of Scotland, that beauty followed a simple trend from north to south, with the highest proportion of uglies in Aberdeen and the greatest frequency of lovelies in London.

Some fashions (tongue piercings, perhaps?) flower once and then disappear, hopefully forever. Others swing in and out of style, as if fastened to the end of a pendulum. Two foibles of human life strongly promote this oscillatory mode. First, our need to create order in a complex world begets our worst mental habit: dichotomy, or our tendency to reduce an intricate set of subtle shadings to a choice between two diametrically opposed alternatives (each with moral weight and therefore ripe for bombast and pontification, if not outright warfare): religion versus science, liberal versus conservative, plain versus fancy, *Roll Over Beethoven* versus the *Moonlight Sonata*. Second, many deep questions about our livelihoods, and the fates of nations, truly have no answers—so we cycle the presumed alternatives of our dichotomies, one after the other, always hoping that, this time, we will find the nonexistent key.

Among oscillating fashions governed primarily by the swing of our social pendulum, no issue could be more prominent for an evolutionary biologist, or more central to a broad range of political questions, than genetic versus environmental sources of human abilities and behaviors. This issue has been falsely dichotomized for so many centuries that English even features a mellifluous linguistic contrast for the supposed alternatives: nature versus nurture.

As any thoughtful person understands, the framing of this question as an either-or dichotomy verges on the nonsensical. Both inheritance and upbringing matter in crucial ways. Moreover, an adult human being, built by interaction of these (and other) factors, cannot be disaggregated into separate components with attached percentages. It behooves us all to grasp why such common claims as "intelligence is 30 percent genetic and 70 percent environmental" have no sensible meaning at all and represent the same kind of error as the contention that all overt properties of water may be revealed by noting an underlying construction from two parts of one gas mixed with one part of another.

Nonetheless, a preference for either nature or nurture swings back and forth into fashion as political winds blow and as scientific breakthroughs grant transient prominence to one or another feature in a spectrum of vital influences. For example, a combination of political and scientific factors favored an emphasis upon environment in the years just following World War II: an understanding that Hitlerian horrors had been rationalized by claptrap genetic theories about inferior races; the domination of psychology by behaviorist theories. Today, genetic explanations are all the rage, fostered by a similar mixture of social and scientific influences: for example, the rightward shift of the political pendulum (and the cynical availability of "you can't change them, they're made that way" as a bogus argument for reducing expenditures on social programs) and an overextension to all behavioral variation of genuinely exciting results in identifying the genetic basis of specific diseases, both physical and mental.

Unfortunately, in the heat of immediate enthusiasm, we often mistake transient fashion for permanent enlightenment. Thus, many people assume that the current popularity of genetic explanation represents a final truth wrested from the clutches of benighted environmental determinists of previous generations. But the lessons of history suggest that the worm will soon turn again. Since both nature and nurture can teach us so much—and since the fullness of our behavior and mentality represents such a complex and unbreakable combination of these and other factors—a current emphasis on nature will no doubt yield to a future fascination with nurture as we move toward better understanding by lurching upward from one side to another in our quest to fulfill the Socratic injunction: know thyself.

In my Galtonian desire to measure the extent of current fascination with genetic explanations (before the pendulum swings once again and my opportunity evaporates), I hasten to invoke two highly newsworthy items of recent months. The subjects may seem quite unrelated—Dolly, the cloned sheep, and Frank Sulloway's book on the effects of birth order upon human behavior—but both stories share a common feature offering striking insight into the current extent of genetic preferences. In short, both stories have been reported almost entirely in genetic terms, but both cry out (at least to me) for a reading as proof of strong environmental influences. Yet no one seems to be drawing (or even mentioning) this glaringly obvious inference. I cannot imagine that anything beyond current fashion for genetic arguments can explain this puzzling silence. I am convinced that exactly the same information, if presented twenty years ago in a climate favoring explanations based on nurture, would have been read primarily in this opposite light. Our world, beset by ignorance and human nastiness, contains quite enough background darkness. Should we not let both beacons shine all the time?

Dolly must be the most famous sheep since John the Baptist designated Jesus in metaphor as "Lamb of God, which taketh away the sin of the world" (John: 1:29). She has certainly edged past the pope, the president, Madonna, and Michael Jordan as the best-known mammal of the moment. And all this for a carbon copy, a Xerox! I don't intent to drip cold water on this little lamb, cloned from a mammary cell of her mother, but I remain unsure that she's worth all the fuss and fear generated by her unconventional birth.

When one reads the technical article describing Dolly's manufacture ("Viable Offspring Derived from Fetal and Adult Mammalian Cells," by I. Wilmut, A. E. Schnieke, J. McWhir, A. J. Kind, and K. H. S. Campbell, *Nature*, February 27, 1997), rather than the fumings and hyperbole of so much public commentary, one can't help feeling a bit underwhelmed and left wondering whether Dolly's story tells less than meets the eye.

I don't mean to discount or underplay the ethical issues raised by Dolly's birth (and I shall return to this subject in a moment), but we are not about to face an army of Hitlers or even a Kentucky Derby run entirely by genetically identical contestants (a true test for the skills of jockeys and trainers). First, Dolly breaks no theoretical ground in biology, for we have known how to clone in principle for at least two decades, but had developed no techniques for reviving the full genetic potential of differentiated adult cells. (Still, I admit that a technological solution can pack as much practical and ethical punch as a theoretical breakthrough. I suppose one could argue that the first atomic bomb only realized a known possibility.)

Second, my colleagues have been able to clone animals from embryonic cell-lines for several years, so Dolly is not the first mammalian clone, but only the first clone from an adult cell. Wilmut and colleagues also cloned sheep from cells of a nine-day embryo and a twenty-six day fetus—and had much greater success. They achieved fifteen pregnancies (although not all proceeded to term) in thirty-two recipients (that is, surrogate mothers for transported cells) of the embryonic cell-line, five pregnancies in sixteen recipients of the fetal cell-line, but only Dolly (one pregnancy in thirteen tries) for the adult cell-line. This experiment cries out for confirming repetition. (Still, I allow that current difficulties will surely be overcome, and cloning from adult cells, if doable at all, will no doubt be achieved more routinely as techniques and familiarity improve.)

Third, and more seriously, I remain unconvinced that we should regard Dolly's starting cell as adult in the usual sense of the term. Dolly grew from a cell taken from the "mammary gland of a six year-old ewe in the last trimester of pregnancy" (to quote the technical article of Wilmut, et al.). Since the breasts of pregnant mammals enlarge substantially in late stages of pregnancy, some mammary cells, although technically adult, may remain unusually labile or even "embryolike" and thus able to proliferate rapidly to produce new breast tissue at an appropriate stage of pregnancy. Consequently, we may be able to clone only from unusual adult cells with effectively embryonic potential, and not from any stray cheek cell, hair follicle, or drop of blood that happens to fall into the clutches of a mad Xeroxer. Wilmut and colleagues admit this possibility in a sentence written with all the obtuseness of conventional scientific prose, and therefore almost universally missed by journalists: "We cannot exclude the possibility that there is a small proportion of relatively undifferentiated stem cells able to support regeneration of the mammary gland during pregnancy."

But if I remain relatively unimpressed by achievements thus far, I do not discount the monumental ethical issues raised by the possibility of cloning from adult cells. Yes, we have cloned fruit trees for decades by the ordinary process of grafting—and without raising any moral alarms. Yes, we may not face the evolutionary dangers of genetic uniformity in crop plants and livestock, for I trust that plant and animal breeders will not be stupid enough to eliminate all but one genotype from a species and will always maintain (as plant breeders do now) an active pool of genetic diversity in reserve. (But then, I suppose we should never underestimate the potential extent of human stupidity—and agricultural seed banks could be destroyed by local catastrophes, while genetic diversity spread throughout a species guarantees maximal evolutionary robustness.)

Nonetheless, while I regard many widely expressed fears as exaggerated, I do worry deeply about potential abuses of human cloning, and I do urge a most open and thorough debate on these issues. Each of us can devise a personal worst-case scenario. Somehow, I do not focus upon the specter of a future Hitler making an army of ten million identical robotic killers, for if our society ever reaches a stage in which such an outcome might be realized, we are probably already lost. My thoughts run to localized moral quagmires that we might actually have to face in the next few years (for example, the biotech equivalent of ambulance-chasing slimeballs among lawyers—a hustling little firm that scans the obits for reports of dead children and then goes to grieving parents with the following offer: "So sorry for your loss, but did you save a hair sample? We can make you another for a mere fifty thou").

However, and still on the subject of ethical conundrums, but now moving to my main point about current underplaying of environmental sources for human behaviors, I do think that the most potent scenarios of fear, and the most fretful ethical discussions on late-night television, have focused on a nonexistent problem that all human societies solved millennia ago. We ask: Is a clone an individual? Would a clone have a soul? Would a clone made from my cell negate my unique personhood?

May I suggest that these endless questions—all variations on the theme that clones threaten our traditional concept of individuality—have already been answered empirically, even though public discussion of Dolly seems blithely oblivious to this evident fact. We have known human clones from the dawn of our consciousness. We call them identical twins—and they are far better clones than Dolly and her mother. Dolly shares only nuclear DNA with her mother's mammary cell, for the nucleus of this cell was inserted into an embryonic stem cell (whose own nucleus had been removed) of a surrogate female. Dolly then grew in the womb of this surrogate.

Identical twins share at least four additional (and important) properties that differ between Dolly and her mother. First, identical twins also house the same mitochondrial genes. (Mitochondria, the "energy factories" of cells, contain a small number of genes. We get our mitochondria from the cytoplasm of the egg cell that made us, not from the nucleus formed by the union of sperm and egg. Dolly received her nucleus from her mother, but her egg cytoplasm, and hence her mitochondria, from her surrogate.) Second, identical twins share the same set of maternal gene products in the egg. Genes don't grow embryos all by themselves. Egg cells contain protein products of maternal genes that play a major role in directing the early development of the embryo. Dolly has her mother's nuclear genes, but her surrogate's gene products in the cytoplasm of her founding cell.

Third—and now we come to explicitly environmental factors—identical twins share the same womb. Dolly and her mother gestated in different places. Fourth, identical twins share the same time and culture (even if they fall into the rare category, so cherished by researchers, of siblings separated at birth and raised, unbeknownst to each other, in distant families of different social classes). The clone of an adult cell matures in a different world. Does anyone seriously believe that a clone of Beethoven would sit down one day to write a Tenth Symphony in the style of his early-nineteenth-century forebear?

So identical twins are truly eerie clones—ever so much more alike on all counts than Dolly and her mother. We do know that identical twins share massive similarities not only of appearance but also in broad propensities and detailed quirks of personality. Nonetheless, have we ever doubted the personhood of each member in a pair of identical twins? Of course not. We know that identical twins are distinct individuals, albeit with peculiar and extensive similarities. We give them different names. They encounter divergent experiences and fates. Their lives wander along disparate paths of the world's complex vagaries. They grow up as distinctive and undoubted individuals, yet they stand forth as far better clones than Dolly and her mother.

Why have we overlooked this central principle in our fears about Dolly? Identical twins provide sturdy proof that inevitable differences of nurture guarantee the individuality and personhood of each human clone. And since any future human Dolly must differ far more from her progenitor (in both the nature of mitochondria and maternal gene products and the nurture of different wombs and surrounding cultures) than any identical twin diverges from her sibling clone, why ask if Dolly has a soul or an independent life when we have never doubted the personhood or individuality of much more similar identical twins?

Literature has always recognized this principle. The Nazi loyalists who cloned Hitler in *The Boys from Brazil* also understood that they had to maximize similarities of nurture as well. So they fostered their little Hitler babies in families maximally like Adolf's own dysfunctional clan—and not one of them grew up anything like history's quintessential monster. Life, too, has always verified this principle. Eng and Chang, the original Siamese twins and the closest clones of all, developed distinct and divergent personalities. One became a morose alcoholic, the other remained a benign and cheerful man. We may not think much of the individuality of sheep in general (for they do set our icon of blind following and identical form as they jump over fences in mental schemes of insomniacs), but Dolly will grow up to be as unique and as ornery as any sheep can be.

A recent book by my friend Frank Sulloway also focuses on themes of nature and nurture. He fretted over, massaged, and lovingly shepherded it toward publication for more than two decades. *Born to Rebel* documents a crucial effect of birth order in shaping human personalities and styles of thinking. Firstborns, as sole recipients of parental attention until the arrival of later children, and as more powerful (by virtue of age and size) than their subsequent siblings, tend to cast their lot with parental authority and with the advantages of incumbent strength. They tend to grow up competent and confident, but also conservative and unlikely to favor quirkiness or innovation. Why threaten an existing structure that has always offered you clear advantages over siblings? Later children, however, are (as Sulloway's title proclaims) born to rebel. They must compete against odds for parental attention long focused primarily elsewhere. They must scrap and struggle and learn to make do for themselves. Laterborns therefore tend to be flexible, innovative, and open to change. The business and political leaders of stable nations may be overwhelmingly firstborns, but the revolutionaries who have discombobulated our cultures and restricted our scientific knowledge tend to be laterborns. Frank and I have been discussing his thesis ever since he began his studies. I thought (and suggested) that he should have published his results twenty years ago. I still hold this opinion, for while I greatly admire his book and do recognize that such a long gestation allowed Frank to strengthen his case by gathering and refining his data, I also believe that he became too committed to his central thesis and tried to extend his explanatory umbrella over too wide a range, with arguments that sometimes smack of special pleading and tortured logic.

Sulloway defends his thesis with statistical data on the relationship of birth order and professional achievement in modern societies—and by interpreting historical patterns as strongly influenced by characteristic differences in behavior of firstborns and laterborns. I found some of his historical arguments fascinating and persuasive when applied to large samples but often uncomfortably overinterpreted in attempts to explain the intricate details of individual lives (for example, the effect of birth order on the differential success of Henry VIII's various wives in overcoming his capricious cruelties).

In a fascinating case, Sulloway chronicles a consistent shift in relative percentages of firstborns among successive groups in power during the French Revolution. The moderates initially in charge tended to be firstborns. As the revolution became more radical, but still idealistic and open to innovation and free discussion, laterborns strongly predominated. But when control then passed to the uncompromising hardliners who promulgated the Reign of Terror, firstborns again ruled the roost. In a brilliant stroke, Sulloway tabulates the birth orders for several hundred delegates who decided the fate of Louis XVI in the National Convention. Among hardliners who voted for the guillotine, 73 percent were firstborns; but of those who opted for the compromise of conviction with pardon, 62 percent were laterborns. Since Louis lost his head by a margin of one vote, an ever so slightly different mix of birth orders among delegates might have altered the course of history.

Since Frank is a good friend and since I have been at least a minor midwife to this project over two decades (although I don't accept all details of his thesis), I took an unusually strong interest in the delayed birth of *Born to Rebel.* I read the text and all the prominent reviews that appeared in many newspapers and journals. And I have been puzzled—stunned would not be too strong a word—by the total absence from all commentary of the simplest and most evident inference from Frank's data, the one glaringly obvious point that everyone should have stressed, given the long history of issues raised by such information.

Sulloway focuses nearly all his interpretation on an extended analogy (broadly valid in my judgement, but overextended as an exclusive device) between birth order in families and ecological status in a world of Darwinian competition. Children vie for limited parental resources, just as individuals struggle for existence (and ultimately for reproductive success) in nature. Birth orders place children in different "niches," requiring disparate modes of competition for maximal success. While firstborns shore up incumbent advantages, laterborns must grope and grub by all clever means at their disposal—leading to the divergent personalities of stalwart and rebel. Alan Wolfe, in my favorite negative review of Sulloway's book from the *New Republic* (December 23, 1996) writes: "Since firstborns already occupy their own niches, laterborns, if they are to be noticed, have to find unoccupied niches. If they do so successfully, they will be rewarded with parental investment." (Jared Diamond stresses the same theme in my favorite positive review from the *New York Review of Books*, November 14, 1996.)

As I said, I am willing to go with this program up to a point. But I must also note that the restriction of commentary to this Darwinian metaphor has diverted attention from the foremost conclusion revealed by a large effect of birth order upon human behavior. The Darwinian metaphor smacks of biology; we also erroneously think of biological explanations as intrinsically genetic (an analysis of this common fallacy could fill an essay or an entire book). I suppose that this chain of argument leads us to stress whatever we think that Sulloway's thesis might be teaching us about "nature" (our preference, in any case, during this age of transient fashion for genetic causes) under our erroneous tendency to treat the explanation of human behavior as a debate between nature and nurture.

But consider the meaning of birth-order effects for environmental influences, however unfashionable at the moment. Siblings differ genetically of course, but no aspect of this genetic variation correlates in any systematic way with birth order. Firstborns and laterborns receive the same genetic shake within a family. Systematic differences in behavior between firstborns and laterborns cannot be ascribed to genetics. (Other biological effects may correlate with birth order—if, for example, the environment of the womb changes systematically with numbers of pregnancies—but such putative influences have no basis in genetic differences among siblings.) Sulloway's substantial birth-order effects therefore provide our best and ultimate documentation of nurture's power. If birth order looms so large in setting the paths of history and the allocation of people to professions, then nurture cannot be denied a powerfully formative role in our intellectual and behavioral variation. To be sure, we often fail to see what stares us in the face, but how can the winds of fashion blow away such an obvious point, one so relevant to our deepest and most persistent questions about ourselves?

In this case, I am especially struck by the irony of fashion's veil. As noted before, I urged Sulloway to publish this data twenty years ago, when (in my judgment) he could have presented an even better case because he had already documented the strong and general influence of birth order upon personality, but had not yet ventured upon the slippery path of trying to explain too many details with forced arguments that sometimes lapse into self-parody. If Sulloway had published in the mid-1970s, when nurture rode the pendulum of fashion in a politically more liberal age (probably dominated by laterborns!), I am confident that this obvious point about birth-order effects as proof of nurture's power would have won primary attention, rather than consignment to a limbo of invisibility.

Hardly anything in intellectual life can be more salutory than the separation of fashion from fact. Always suspect fashion (especially when the moment's custom matches your personal predilection); always cherish fact (while remembering than an apparent "fact" may only record a transient fashion). I have discussed two subjects that couldn't be "hotter," but cannot be adequately understood because a veil of genetic fashion now conceals the richness of full explanation by relegating a preeminent environmental theme to invisibility. Thus, we worry whether the first cloned sheep represents a genuine individual at all, while we forget that we have never doubted the distinct personhood guaranteed by differences in nurture to clones far more similar by nature than Dolly and her mother—identical twins. And we try to explain the strong effects of birth order only by invoking a Darwinian analogy between family place and ecological niche, while forgetting that these systematic effects cannot have a genetic basis and therefore prove the predictable power of nurture.

So sorry, Louis. You lost your head to the power of family environments upon head children. And hello, Dolly. May we forever restrict your mode of manufacture, at least for humans. But may genetic custom never stale the infinite variety guaranteed by a lifetime of nurture in the intricate complexity of nature—this vale of tears, joy, and endless wonder.

The Wisdom of Repugnance

by Leon R. Kass

Leon R. Kass is Addie Clark Harding Professor in the College and the Committee on Social Thought, University of Chicago.

OUR HABIT OF delighting in news of scientific and technological breakthroughs has been sorely challenged in the birth announcement of a sheep named Dolly. Though Dolly shares with previous sheep the "softest clothing, woolly, bright," William Blake's question, "Little Lamb, who made thee?" has for her a radically different answer: Dolly was, quite literally, made. She is the work not of nature or nature's God but of man, an Englishman, Ian Wilmut, and his fellow scientists. What's more, Dolly came into being not only asexually—ironically, just like "He [who] calls Himself a Lamb"—but also as the genetically identical copy (and the perfect incarnation of the form or blueprint) of a mature ewe, of whom she is a clone. This long-awaited yet not quite expected success in cloning a mammal raised immediately the prospect—and the specter—of cloning human beings: "I a child and Thou a lamb," despite our differences, have always been equal candidates for creative making, only now, by means of cloning, we may both spring from the hand of man playing at being God.

After an initial flurry of expert comment and public consternation, with opinion polls showing overwhelming opposition to cloning human beings, President Clinton ordered a ban on all federal support for human cloning research (even though none was being supported) and charged the National

---

"The Wisdom of Repugnance: Why We Should Ban the Cloning of Humans," by Leon R. Kass, *New Republic*, vol. 2 (June 1997): 17–26. Reprinted with permission from Leon R. Kass.

Bioethics Advisory Commission to report in ninety days on the ethics of human cloning research. The commission (an eighteen-member panel, evenly balanced between scientists and nonscientists appointed by the president and reporting to the National Science and Technology Council) invited testimony from scientists, religious thinkers and bioethicists, as well as from the general public. It is now deliberating about what it should recommend, both as a matter of ethics and as a matter of public policy.

Congress is awaiting the commission's report, and is poised to act. Bills to prohibit the use of federal funds for human cloning research have been introduced in the House of Representatives and the Senate; and another bill, in the House, would make it illegal "for any person to use a human somatic cell for the process of producing a human clone." A fateful decision is at hand. To clone or not to clone a human being is no longer an academic question.

Taking Cloning Seriously, Then and Now

Cloning first came to public attention roughly thirty years ago, following the successful asexual production, in England, of a clutch of tadpole clones by the technique of nuclear transplantation. The individual largely responsible for bringing the prospect and promise of human cloning to public notice was Joshua Lederberg, a Nobel Laureate geneticist and a man of large vision. In 1966, Lederberg wrote a remarkable article in *The American Naturalist* detailing the eugenic advantages of human cloning and other forms of genetic engineering, and the following year he devoted a column in *The Washington Post*, where he wrote regularly on science and society, to the prospect of human cloning. He suggested that cloning could help us overcome the unpredictable variety that still rules human reproduction, and allow us to benefit from perpetuating superior genetic endowments. These writings sparked a small public debate in which I became a participant. At the time a young researcher in molecular biology at the National Institutes of Health (NIH), I wrote a reply to the *Post*, arguing against Lederberg's amoral treatment of this morally weighty subject and insisting on the urgency of confronting a series of questions and objections, culminating in the suggestion that "the programmed reproduction of man will, in fact, dehumanize him."

Much has happened in the intervening years. It has become harder, not easier, to discern the true meaning of human cloning. We have in some sense been softened up to the idea—through movies, cartoons, jokes and intermittent commentary in the mass media, some serious, most lighthearted. We have become accustomed to new practices in human reproduction: not just in vitro fertilization, but also embryo manipulation, embryo donation and surrogate pregnancy. Animal biotechnology has yielded transgenic animals and a burgeoning science of genetic engineering, easily and soon to be transferable to humans.

Even more important, changes in the broader culture make it now vastly more difficult to express a common and respectful understanding of sexuality, procreation, nascent life, family, and the meaning of motherhood, fatherhood and the links between the generations. Twenty-five years ago, abortion was still largely illegal and thought to be immoral, the sexual revolution (made possible by the extramarital use of the pill) was still in its infancy, and few had yet heard about the reproductive rights of single women, homosexual men and lesbians. (Never mind shameless memories about one's own incest!) Then one could argue, without embarrassment, that the new technologies of human reproduction—babies without sex—and their confounding of normal kin relations—who's the mother: the egg donor, the surrogate who carries and delivers, or the one who rears?—would "undermine the justification and support that biological parenthood gives to the monogamous marriage." Today, defenders of stable, monogamous marriage risk charges of giving offense to those adults who are living in "new family forms" or to those children who, even without the benefit of assisted reproduction, have acquired either three or four parents or one or none at all. Today, one must even apologize for voicing opinions that twenty-five years ago were nearly universally regarded as the core of our culture's wisdom on these matters. In a world whose once-given natural boundaries are blurred by technological change and whose moral boundaries are seemingly up for grabs, it is much more difficult to make persuasive the still compelling case against cloning human beings. As Raskolnikov put it, "man gets used to everything—the beast!"

Indeed, perhaps the most depressing feature of the discussions that immediately followed the news about Dolly was their ironical tone, their genial cynicism, their moral fatigue: "AN UDDER WAY OF MAKING LAMBS" (*Nature*), "WHO WILL CASH IN ON BREAKTHROUGH IN CLONING?" (*The Wall Street Journal*), "IS CLONING BAAAAAAAD?" (*The Chicago Tribune*). Gone from the scene are the wise and courageous voices of Theodosius Dobzhansky (genetics), Hans Jonas (philosophy) and Paul Ramsey (theology) who, only twenty-five years ago, all made powerful moral arguments against ever cloning a human being. We are now too sophisticated for such argumentation: we wouldn't be caught in public with a strong moral stance, never mind an absolutist one. We are all, or almost all, post-modernists now.

Cloning turns out to be the perfect embodiment of the ruling opinions of our new age. Thanks to the sexual revolution, we are able to deny in practice, and increasingly in thought, the inherent procreative teleology of sexuality itself. But, if sex has no intrinsic connection to generating babies, babies need have no necessary connection to sex. Thanks to feminism and the gay rights movement, we are increasingly encouraged to treat the natural heterosexual difference and its preeminence as a matter of "cultural construction." But if male and female are not normatively complementary and generatively significant, babies need not come from male and female complementarily. Thanks to the prominence and the acceptability of divorce and out-of-wedlock births, stable, monogamous marriage as the ideal home for procreation is no longer the agreed-upon cultural norm. For this new dispensation, the clone is the ideal emblem: the ultimate "single-parent child."

Thanks to our belief that all children should be *wanted* children (the more high-minded principle we use to justify contraception and abortion), sooner or later only those children who fulfill our wants will be fully acceptable. Through cloning, we can work our wants and wills on the very identity of our children, exercising control as never before. Thanks to modern notions of individualism, and the rate of cultural change, we see ourselves not as linked to ancestors and defined by traditions, but as projects for our own self-creation, not only as self-made men but also man-made selves; and self-cloning is simply an extension of such rootless and narcissistic self-re-creation.

Unwilling to acknowledge our debt to the past and unwilling to embrace the uncertainties and the limitations of the future, we have a false relation to both: cloning personifies our desire fully to control the future, while being subject to no controls ourselves. Enchanted and enslaved by the glamour of technology, we have lost our awe and wonder before the deep mysteries of nature and of life. We cheerfully take our own beginnings in our hands and, like the last man, we blink.

Part of the blame for our complacency lies, sadly, with the field of bioethics itself, and its claim to expertise in these moral matters. Bioethics was founded by people who understood that the new biology touched and threatened the deepest matters of our humanity: bodily integrity, identity and individuality, lineage and kinship, freedom and self-command, eros and aspiration, and the relations and strivings of body and soul. With its capture by analytic philosophy, however, and its inevitable routinization and professionalization, the field has by and large come to content itself with analyzing moral arguments, reacting to new technological developments and taking on emerging issues of public policy, all performed with a naive faith that the evils we fear can all be avoided by compassion, regulation and a respect for autonomy. Bioethics has made some major contributions in the protection of human subjects and in other areas where personal freedom is threatened: but its practitioners, with few exceptions, have turned the big human questions into pretty thin gruel.

One reason for this is that the piecemeal formation of public policy tends to grind down large questions of morals into small questions of procedure. Many of the country's leading bioethicists have served on national commissions or state task forces and advisory boards, where, understandably, they have found utilitarianism to be the only ethical vocabulary acceptable to all participants in discussing issues of law, regulation and public policy. As many of these commissions have been either officially under the aegis of NIH or the Health and Human Services Department, or otherwise dominated by powerful voices for scientific progress, the ethicists have for the most part been content, after some "values clarification" and wringing of hands, to pronounce their blessings upon the inevitable. Indeed, it is the bioethicists, not the scientists, who are now the most articulate defenders of human cloning: the two witnesses testifying before the National Bioethics Advisory Commission in favor of cloning human beings were bioethicists, eager to rebut what they regard as the irrational concerns of those of us in opposition. One wonders whether this commission, constituted like the previous commissions, can tear itself sufficiently free from the accommodationist pattern of rubber-stamping all technical innovation, in the mistaken belief that all other goods must bow down before the gods of better health and scientific advance.

If it is to do so, the commission must first persuade itself, as we all should persuade ourselves, not to be complacent about what is at issue here. Human cloning, though it is in some respects continuous with previous reproductive technologies, also represents something radically new, in itself and in its easily foreseeable consequences. The stakes are very high indeed. I exaggerate, but in the direction of the truth, when I insist that we are faced with having to decide nothing less than whether human procreation is going to remain human, whether children are going to be made rather than begotten, whether it is a good thing, humanly speaking, to say yes in principle to the road which leads (at best) to the dehumanized rationality of *Brave New World*. This is not business as usual, to be fretted about for a while but finally to be given our seal of approval. We must rise to the occasion and make our judgments as if the future of our humanity hangs in the balance. For so it does.

The State of the Art

If we should not underestimate the significance of human cloning, neither should we exaggerate its imminence or misunderstand just what is involved. The procedure is conceptually simple. The nucleus of a mature but unfertilized egg is removed and replaced with a nucleus obtained from a specialized cell of an adult (or fetal) organism (in Dolly's case, the donor nucleus came from mammary gland epithelium). Since almost all the hereditary material of a cell is contained within its nucleus, the renucleated egg and the individual into which this egg develops are genetically identical to the organism that was the source of the transferred nucleus. An unlimited number of genetically identical individuals—clones—could be produced by nuclear transfer. In principle, any person, male or female, newborn or adult, could be cloned, and in any quantity. With laboratory cultivation and storage of tissues, cells outliving their sources make it possible even to clone the dead.

The technical stumbling block, overcome by Wilmut and his colleagues, was to find a means of reprogramming the state of the DNA in the donor cells, reversing its differentiated expression and restoring its full totipotency, so that it could again direct the entire process of producing a mature organism. Now that this problem has been solved, we should expect a rush to develop cloning for other animals, especially livestock, in order to propagate in perpetuity the champion meat or milk producers. Though exactly how soon someone will succeed in cloning a human being is anybody's guess, Wilmut's technique, almost certainly applicable to humans, makes *attempting* the feat an imminent possibility.

Yet some cautions are in order and some possible misconceptions need correcting. For a start, cloning is not Xeroxing. As has been reassuringly reiterated, the clone of Mel Gibson, though his genetic double, would enter the world hairless, toothless and peeing in his diapers, just like any other human infant. Moreover, the success rate, at least at first, will probably not be very high: the British transferred 277 adult nuclei into enucleated sheep eggs, and implanted twenty-nine clonal embryos, but they achieved the birth of only one live lamb clone. For this reason, among others, it is unlikely that, at least for now, the practice would be very popular, and there is no immediate worry of mass-scale production of multicopies. The need of repeated surgery to obtain eggs and, more crucially, of numerous borrowed wombs for implantation will surely limit use, as will the expense; besides, almost everyone who is able will doubtless prefer nature's sexier way of conceiving.

Still, for the tens of thousands of people already sustaining over 200 assisted-reproduction clinics in the United States and already availing themselves of in vitro fertilization, intracytoplasmic sperm injection and other techniques of assisted reproduction, cloning would be an option with virtually no added fuss (especially, when the success rate improves). Should commercial interests develop in "nucleus-banking," as they have in sperm-banking; should famous athletes or other celebrities decide to market their DNA the way they now market their autographs and just about everything else; should techniques of embryo and germline genetic testing and manipulation arrive as anticipated, increasing the use of laboratory assistance in order to obtain "better" babies—should all this come to pass, then cloning, if it is permitted, could become more than a marginal practice simply on the basis of free reproductive choice, even without any social encouragement to upgrade the gene pool or to replicate superior types. Moreover, if laboratory research on human cloning proceeds, even without any intention to produce cloned humans, the existence of cloned human embryos in the laboratory, created to begin with only for research purposes, would surely pave the way for later baby-making implantations.

In anticipation of human cloning, apologists and proponents have already made clear possible uses of the perfected technology, ranging from the sentimental and compassionate to the grandiose. They include: providing a child for an infertile couple; "replacing" a beloved spouse or child who is dying or had died; avoiding the risk of genetic disease; permitting reproduction for homosexual men and lesbians who want nothing sexual to do with the opposite sex; securing a genetically identical source of organs or tissues perfectly suitable for transplantation; getting a child with a genotype of one's own choosing, not excluding oneself; replicating individuals of great genius, talent or beauty—having a child who really could "be like Mike"; and creating large sets of genetically identical humans suitable for research on, for instance, the question of nature versus nurture, or for special missions in peace and war (not excluding espionage), in which using identical humans would be an advantage. Most people who envision the cloning of human beings, of course, want none of these scenarios. That they cannot say why is not surprising. What is surprising, and welcome, is that, in our cynical age, they are saying anything at all.

The Wisdom of Repugnance

"Offensive." "Grotesque." "Revolting." "Repugnant." "Repulsive." These are the words most commonly heard regarding the prospect of human cloning. Such reactions come both from the man or woman in the street and from the intellectuals, from believers and atheists, from humanists and scientists. Even Dolly's creator has said he "would find it offensive" to clone a human being.

People are repelled by many aspects of human cloning. They recoil from the prospect of mass production of human beings, with large clones of look-alikes, compromised in their individuality; the idea of father-son or mother-daughter twins; the bizarre prospects of a woman giving birth to and rearing a genetic copy of herself, her spouse or even her deceased father or mother; the grotesqueness of conceiving a child as an exact replacement for another who has died; the utilitarian creation of embryonic genetic duplicates of oneself, to be frozen away or created when necessary, in case of need for homologous tissues or organs for transplantation; the narcissism of those who would clone themselves and the arrogance of others who think they know who deserves to be cloned or which genotype any child-to-be should be thrilled to receive; the Frankensteinian hubris to create human life and increasingly to control its destiny; man playing God. Almost no one finds any of the suggested reasons for human cloning compelling; almost everyone anticipates its possible misuses and abuses. Moreover, many people feel oppressed by the sense that there is probably nothing we can do to prevent it from happening. This makes the prospect all the more revolting.

Revulsion is not an argument; and some of yesterday's repugnances are today calmly accepted—though, one must add, not always for the better. In crucial cases, however, repugnance is the emotional expression of deep wisdom, beyond reason's power fully to articulate it. Can anyone really give an argument fully adequate to the horror which is father-daughter incest (even with consent), or having sex with animals, or mutilating a corpse, or eating human flesh, or even just (just!) raping or murdering another human being? Would anybody's failure to give full rational justification for his or her revulsion at these practices make that revulsion ethically suspect? Not at all. On the contrary, we are suspicious of those who think that they can rationalize away our horror, say, by trying to explain the enormity of incest with arguments only about the genetic risks of inbreeding.

The repugnance at human cloning belongs in this category. We are repelled by the prospect of cloning human beings not because of the strangeness or novelty of the undertaking, but because we intuit and feel, immediately and without arguments, the violation of things that we rightfully hold dear. Repugnance, here as elsewhere, revolts against the excesses of human willfulness, warning us not to transgress what is unspeakably profound. Indeed, in this age in which everything is held to be permissible so long as it is freely done, in which our given human nature no longer commands respect, in which our bodies are regarded as mere instruments of our autonomous rational wills, repugnance may be the only voice left that speaks up to defend the central core of our humanity. Shallow are the souls that have forgotten how to shudder.

The goods protected by repugnance are generally overlooked by our customary ways of approaching all new biomedical technologies. The way we evaluate cloning ethically will in fact be shaped by how we characterize it descriptively, by the context into which we place it, and by the perspective from which we view it. The first task for ethics is proper description. And here is where our failure begins.

Typically, cloning is discussed in one or more of three familiar contexts, which one might call the technological, the liberal and the meliorist. Under the first, cloning will be seen as an extension of existing techniques for assisting reproduction and determining the genetic makeup of children. Like them, cloning is to be regarded as a neutral technique, with no inherent meaning or goodness, but subject to multiple uses, some good, some bad. The morality of cloning thus depends absolutely on the goodness or badness of the motives and intentions of the cloners: as one bioethicist defender of cloning puts it, "the ethics must be judged [only] by the way the parents nurture and rear their resulting child and whether they bestow the same love and affection on a child brought into existence by a technique of assisted reproduction as they would on a child born in the usual way."

The liberal (or libertarian or liberationist) perspective sets cloning in the context of rights, freedoms and personal empowerment. Cloning is just a new option for exercising an individual's right to reproduce or to have the kind of child that he or she wants. Alternatively, cloning enhances our liberation (especially women's liberation) from the confines of nature, the vagaries of chance, or the necessity for sexual mating. Indeed, it liberates women from the need for men altogether, for the process requires only eggs, nuclei and (for the time being) uteri—plus, of course, a healthy dose of our (allegedly "masculine") manipulative science that likes to do all these things to mother nature and nature's mothers. For those who hold this outlook, the only moral restraints on cloning are adequately informed consent and the avoidance of bodily harm. If no one is cloned without her consent, and if the clonant is not physically damaged, then the liberal conditions for licit, hence moral, conduct are met. Worries that go beyond violating the will or maiming the body are dismissed as "symbolic"—which is to say, unreal.

The meliorist perspective embraces valetudinarians and also eugenicists. The latter were formerly more vocal in these discussions, but they are now generally happy to see their goals advanced under the less threatening banners of freedom and technological growth. These people see in cloning a new prospect for improving human beings—minimally, by ensuring the perpetuation of healthy individuals by avoiding the risks of genetic disease inherent in the lottery of sex, and maximally, by producing "optimum babies," preserving outstanding genetic material, and (with the help of soon-to-come techniques for precise genetic engineering) enhancing inborn human capacities on many fronts. Here the morality of cloning as a means is justified solely by the excellence of the end, that is, by the outstanding traits or individuals cloned—beauty, or brawn, or brains.

These three approaches, all quintessentially American and all perfectly fine in their places, are sorely wanting as approaches to human procreation. It is, to say the least, grossly distorting to view the wondrous mysteries of birth, renewal and individuality, and the deep meaning of parent-child relations, largely through the lens of our reductive science and its potent technologies. Similarly, considering reproduction (and the intimate relations of family life!) primarily under the political-legal, adversarial and individualistic notion of rights can only undermine the private yet fundamentally social, cooperative and duty-laden character of child-bearing, child-rearing and their bond to the covenant of marriage. Seeking to escape entirely from nature (in order to satisfy a natural desire or a natural right to reproduce!) is self-contradictory in theory and self-alienating in practice. For we are erotic beings only because we are embodied beings, and not merely intellects and wills unfortunately imprisoned in our bodies. And, though health and fitness are clearly great goods, there is something deeply disquieting in looking on our prospective children as artful products perfectible by genetic engineering, increasingly held to our willfully imposed designs, specifications and margins of tolerable error.

The technical, liberal and meliorist approaches all ignore the deeper anthropological, social and, indeed, ontological meanings of bringing forth new life. To this more fitting and profound point of view, cloning shows itself to be a major alteration, indeed, a major violation, of our given nature as embodied, gendered and engendering beings—and of the social relations built on this natural ground. Once this perspective is recognized, the ethical judgment on cloning can no longer be reduced to a matter of motives and intentions, rights and freedoms, benefits and harms, or even means and ends. It must be regarded primarily as a matter of meaning: Is cloning a fulfillment of human begetting and belonging? Or is cloning rather, as I contend, their pollution and perversion? To pollution and perversion, the fitting response can only be horror and revulsion: and conversely, generalized horror and revulsion are prima facie evidence of foulness and violation. The burden of moral argument must fall entirely on those who want to declare the widespread repugnances of humankind to be mere timidity or superstition.

Yet repugnance need not stand naked before the bar of reason. The wisdom of our horror at human cloning can be partially articulated, even if this is finally one of those instances about which the heart has its reasons that reason cannot entirely know.

The Profundity of Sex

To see cloning in its proper context, we must begin not, as I did before, with laboratory technique, but with the anthropology—natural and social—of sexual reproduction.

Sexual reproduction—by which I mean the generation of new life from (exactly) two complementary elements, one female, one male, (usually) through coitus—is established (if that is the right term) not by human decision, culture or tradition, but by nature; it is the natural way of all mammalian reproduction. By nature, each child has two complementary biological progenitors. Each child thus stems from and unites exactly two lineages. In natural generation, moreover, the precise genetic constitution of the resulting offspring is determined by a combination of nature and chance, not by human design: each human child shares the common natural human species genotype, each child is genetically (equally) kin to each (both) parent(s), yet each child is also genetically unique.

These biological truths about our origins foretell deep truths about our identity and about our human condition altogether. Every one of us is at once equally human, equally enmeshed in a particular familial nexus of origin, and equally individuated in our trajectory from birth to death—and, if all goes well, equally capable (despite our mortality) of participating, with a complementary other, in the very same renewal of such human possibility through procreation. Though less momentous than our common humanity, our genetic individuality is not humanly trivial. It shows itself forth in our distinctive appearance through which we are everywhere recognized; it is revealed in our "signature" marks of fingerprints and our self-recognizing immune system; it symbolizes and foreshadows exactly the unique, never-to-be-repeated character of each human life.

Human societies virtually everywhere have structured child-rearing responsibilities and systems of identity and relationship on the bases of these deep natural facts of begetting. The mysterious yet ubiquitous "love of one's own" is everywhere culturally exploited, to make sure that children are not just produced but well cared for and to create for everyone clear ties of meaning, belonging and obligation. But it is wrong to treat such naturally rooted social practices as mere cultural constructs (like left- or right-driving, or like burying or cremating the dead) that we can alter with little human cost. What would kinship be without its clear natural grounding? And what would identity be without kinship? We must resist those who have begun to refer to sexual reproduction as the "traditional method of reproduction," who would have us regard as merely traditional, and by implication arbitrary, what is in truth not only natural but most certainly profound.

Asexual reproduction, which produces "single-parent" offspring, is a radical departure from the natural human way, confounding all normal understandings of father, mother, sibling, grandparent, etc., and all moral relations tied thereto. It becomes even more of a radical departure when the resulting offspring is a clone derived not from an embryo, but from a mature adult to whom the clone would be an identical twin; and when the process occurs not by natural accident (as in natural twinning), but by deliberate human design and manipulation; and when the child's (or children's) genetic constitution is pre-selected by the parent(s) (or scientists). Accordingly, as we will see, cloning is vulnerable to three kinds of concerns and objections, related to these three points: cloning threatens confusion of identity and individuality, even in small-scale cloning; cloning represents a giant step (though not the first one) toward transforming procreation into manufacture, that is, toward the increasing depersonalization of the process of generation and, increasingly, toward the "production" of human children as artifacts, products of human will and design (what others have called the problem of "commodification" of new life); and cloning—like other forms of eugenic engineering of the next generation—represents a form of despotism of the cloners over the cloned, and thus (even in benevolent cases) represents a blatant violation of the inner meaning of parent-child relations, of what it means to have a child, of what it means to say "yes" to our own demise and "replacement."

Before turning to these specific ethical objections, let me test my claim of the profundity of the natural way by taking up a challenge recently posed by a friend. What if the given natural human way of reproduction were asexual, and we now had to deal with a new technological innovation—artificially induced sexual dimorphism and the fusing of complementary gametes—whose inventors argued that sexual reproduction promised all sorts of advantages, including hybrid vigor and the creation of greatly increased individuality? Would one then be forced to defend natural asexuality because it was natural? Could one claim that it carried deep human meaning?

The response to this challenge broaches the ontological meaning of sexual reproduction. For it is impossible, I submit, for there to have been human life—or even higher forms of animal life—in the absence of sexuality and sexual reproduction. We find asexual reproduction only in the lowest forms of life: bacteria, algae, fungi, some lower invertebrates. Sexuality brings with it a new and enriched relationship to the world. Only sexual animals can seek and find complementary others with whom to pursue a goal that transcends their own existence. For a sexual being, the world is no longer an indifferent and largely homogeneous *otherness*, in part edible, in part dangerous. It also contains some very special and related and complementary beings, of the same kind but of opposite sex, toward whom one reaches out with special interest and intensity. In higher birds and mammals, the outward gaze keeps a lookout not only for food and predators, but also for prospective mates; the beholding of the many splendored world is suffused with desire for union, the animal antecedent of human eros and the germ of sociality. Not by accident is the human animal both the sexiest animal—whose females do not go into heat but are receptive throughout the estrous cycle and whose males must therefore have greater sexual appetite and energy in order to reproduce successfully—and also the most aspiring, the most social, the most open and the most intelligent animal.

The soul-elevating power of sexuality is, at bottom, rooted in its strange connection to mortality, which it simultaneously accepts and tries to overcome. Asexual production may be seen as a continuation of the activity of self-preservation. When one organism buds or divides to become two, the original being is (doubly) preserved, and nothing dies. Sexuality, by contrast, means perishability and serves replacement: the two that come together to generate one soon will die. Sexual desire, in human beings as in animals, thus serves an end that is partly hidden from, and finally at odds with, the self-serving individual. Whether we know it or not, when we are sexually active we are voting with our genitalia for our own demise. The salmon swimming upstream to spawn and die tell the universal story: sex is bound up with death, to which it holds a partial answer in procreation.

The salmon and the other animals evince this truth blindly. Only the human being can understand what it means. As we learn so powerfully from the story of the Garden of Eden, our humanization is coincident with sexual self-consciousness, with the recognition of our sexual nakedness and all that it implies: shame at our needy incompleteness, unruly self-division and finitude; awe before the eternal; hope in the self-transcending possibilities of children and a relationship to the divine. In the sexually self-conscious animal, sexual desire can become eros, lust can become love. Sexual desire humanly regarded is thus sublimated into erotic longing for wholeness, completion and immortality, which drives us knowingly into the embrace and its generative fruit—as well as into all the higher human possibilities of deed, speech and song.

Through children, a good common to both husband and wife, male and female achieve some genuine unification (beyond the mere sexual "union," which fails to do so). The two become one through sharing generous (not needy) love for this third being as good. Flesh of their flesh, the child is the parents' own commingled being externalized, and given a separate and persisting existence. Unification is enhanced also by their commingled work of rearing. Providing an opening to the future beyond the grave, carrying not only our seed but also our names, our ways and our hopes that they will surpass us in goodness and happiness, children are a testament to the possibility of transcendence. Gender duality and sexual desire, which first draws our love upward and outside of ourselves, finally provide for the partial overcoming of the confinement and limitation of perishable embodiment altogether.

Human procreation, in sum, is not simply an activity of our rational wills. It is a more complete activity precisely because it engages us bodily, erotically and spiritually, as well as rationally. There is wisdom in the mystery of nature that has joined the pleasure of sex, the inarticulate longing for union, the communication of the loving embrace and the deep-seated and only partly articulate desire for children in the very activity by which we continue the chain of human existence and participate in the renewal of human possibility. Whether or not we know it, the severing of procreation from sex, love and intimacy is inherently dehumanizing, no matter how good the product.

We are now ready for the more specific objections to cloning.

The Perversities of Cloning

First, an important if formal objection: any attempt to clone a human being would constitute an unethical experiment upon the resulting child-to-be. As the animal experiments (frog and sheep) indicate, there are grave risks of mishaps and deformities. Moreover, because of what cloning means, one cannot presume a future cloned child's consent to be a clone, even a healthy one. Thus, ethically speaking, we cannot even get to know whether or not human cloning is feasible.

I understand, of course, the philosophical difficulty of trying to compare a life with defects against nonexistence. Several bioethicists, proud of their philosophical cleverness, use this conundrum to embarrass claims that one can injure a child in its conception, precisely because it is only thanks to that complained-of conception that the child is alive to complain. But common sense tells us what we have no reason to fear such philosophisms. For we surely know that people can harm and even maim children in the very act of conceiving them, say, by paternal transmission of the AIDS virus, maternal transmission of heroin dependence or, arguably, even by bringing them into being as bastards or with no capacity or willingness to look after them properly. And we believe that to do this intentionally, or even negligently, is inexcusable and clearly unethical.

The objection about the impossibility of presuming consent may even go beyond the obvious and sufficient point that a clonant, were he subsequently to be asked, could rightly resent having been made a clone. At issue are not just benefits and harms, but doubts about the very independence needed to give proper (even retroactive) consent, that is, not just the capacity to choose but the disposition and ability to chose freely and well. It is not at all clear to what extent a clone will truly be a moral agent. For, as we shall see, in the very fact of cloning, and of rearing him as a clone, his makers subvert the cloned child's independence, beginning with that aspect that comes from knowing that one was an unbidden surprise, a gift, to the world, rather than the designed result of someone's artful project.

Cloning creates serious issues of identity and individuality. The cloned person may experience concerns about his distinctive identity not only because he will be in genotype and appearance identical to another human being, but, in this case, because he may also be twin to the person who is his "father" or "mother"—if one can still call them that. What would be the psychic burdens of being the "child" or "parent" of your twin? The cloned individual, moreover, will be saddled with a genotype that has already lived. He will not be fully a surprise to the world. People are likely always to compare his performances in life with that of his alter ego. True, his nurture and his circumstance in life will be different; genotype is not exactly destiny. Still, one must also expect parental and other efforts to shape this new life after the original—or at least to view the child with the original version always firmly in mind. Why else did they clone from the star basketball player, mathematician and beauty queen—or even dear old dad—in the first place?

Since the birth of Dolly, there has been a fair amount of doublespeak on this matter of genetic identity. Experts have rushed in to reassure the public that the clone would in no way be the same person, or have any confusions about his or her identity: as previously noted, they are pleased to point out that the clone of Mel Gibson would not be Mel Gibson. Fair enough. But one is shortchanging the truth by emphasizing the additional importance of the intrauterine environment, rearing and social setting: genotype obviously matters plenty. That, after all, is the only reason to clone, whether human beings or sheep. The odds that clones of Wilt Chamberlain will play in the NBA are, I submit, infinitely greater than they are for clones of Robert Reich.

Curiously, this conclusion is supported, inadvertently, by the one ethical sticking point insisted on by friends of cloning: no cloning without the donor's consent. Though an orthodox liberal objection, it is in fact quite puzzling when it comes from people (such as Ruth Macklin) who also insist that genotype is not identity or individuality, and who deny that a child could reasonably complain about being made a genetic copy. If the clone of Mel Gibson would not be Mel Gibson, why should Mel Gibson have grounds to object that someone had been made his clone? We already allow researchers to use blood and tissue samples for research purposes of no benefit to their sources: my falling hair, my expectorations, my urine and even my biopsied tissues are "not me" and not mine. Courts have held that the profit gained from uses to which scientists put my discarded tissues do not legally belong to me. Why, then, no cloning without consent—including, I assume, no cloning from the body of someone who just died? What harm is done the donor, if genotype is "not me"? Truth to tell, the only powerful justification for objecting is that genotype really does have something to do with identity, and everybody knows it. If not, on what basis could Michael Jordan object that someone cloned "him," say, from cells taken from a "lost" scraped-off piece of his skin? The insistence on donor consent unwittingly reveals the problem of identity in all cloning.

Genetic distinctiveness not only symbolizes the uniqueness of each human life and the independence of its parents that each human child rightfully attains. It can also be an important support for living a worthy and dignified life. Such arguments apply with great force to any large-scale replication of human individuals. But they are sufficient, in my view, to rebut even the first attempts to clone a human being. One must never forget that these are human beings upon whom our eugenic or merely playful fantasies are to be enacted.

Troubled psychic identity (distinctiveness), based on all-too-evident genetic identity (sameness), will be made much worse by the utter confusion of social identity and kinship ties. For, as already noted, cloning radically confounds lineage and social relations, for "offspring" as for "parents." As bioethicist James Nelson has pointed out, a female child cloned from her "mother" might develop a desire for a relationship to her "father," and might understandably seek out the father of her "mother," who is after all also her biological twin sister. Would "grandpa," who thought his paternal duties concluded, be pleased to discover that the clonant looked to him for paternal attention and support?

Social identity and social ties of relationship and responsibility are widely connected to, and supported by, biological kinship. Social taboos on incest (and adultery) everywhere serve to keep clear who is related to whom (and especially which child belongs to which parents), as well as to avoid confounding the social identity of parent-and-child (or brother-and-sister) with the social identity of lovers, spouses and co-parents. True, social identity is altered by adoption (but as a matter of the best interest of already living children: we do not deliberately produce children for adoption). True, artificial insemination and in vitro fertilization with donor sperm, or whole embryo donation, are in some way forms of "prenatal adoption"—a not altogether unproblematic practice. Even here, though, there is in each case (as in all sexual reproduction) a known male source of sperm and a known single female source of egg—a genetic father and a genetic mother—should anyone care to know (as adopted children often do) who is genetically related to whom.

In the case of cloning, however, there is but one "parent." The usually sad situation of the "single-parent child" is here deliberately planned, and with a vengeance. In the case of self-cloning, the "offspring" is, in addition, one's twin; and so the dreaded result of incest—to be parent to one's sibling—is here brought about deliberately, albeit without any act of coitus. Moreover, all other relationships will be confounded. What will father, grandfather, aunt, cousin, sister mean? Who will bear what ties and what burdens? What sort of social identity will someone have with one whole side—"father's" or "mother's"—necessarily excluded? It is no answer to say that our society, with its high incidence of divorce, remarriage, adoption, extramarital child-bearing and the rest, already confounds lineage and confuses kinship and responsibility for children (and everyone else), unless one also wants to argue that this is, for children, a preferable state of affairs.

Human cloning would also represent a giant step toward turning begetting into making, procreation into manufacture (literally, something "handmade"), a process already begun with in vitro fertilization and genetic testing of embryos. With cloning, not only is the process in hand, but the total genetic blueprint of the cloned individual is selected and determined by the human artisans. To be sure, subsequent development will take place according to natural processes; and the resulting children will still be recognizably human. But we here would be taking a major step into making man himself simply another one of the man-made things. Human nature becomes merely the last part of nature to succumb to the technological project, which turns all of nature into raw material at human disposal, to be homogenized by our rationalized technique according to the subjective prejudices of the day.

How does begetting differ from making? In natural procreation, human beings come together, complementarily male and female, to give existence to another being who is formed, exactly as we were, *by what we are*: living, hence perishable, hence aspiringly erotic, human beings. In clonal reproduction, by contrast, and in the more advanced forms of manufacture to which it leads, we give existence to a being not by what we are but by what we intend and design. As with any product of our making, no matter how excellent, the artificer stands above it, not as an equal but as a superior, transcending it by his will and creative prowess. Scientists who clone animals make it perfectly clear that they are engaged in instrumental making; the animals are, from the start, designed as means to serve rational human purposes. In human cloning, scientist and prospective "parents" would be adopting the same technocratic mentality to human children: human children would be their artifacts.

Such an arrangement is profoundly dehumanizing, no matter how good the product. Mass-scale cloning of the same individual makes the point vividly; but the violation of human equality, freedom and dignity are present even in a single planned clone. And procreation dehumanized into manufacture is further degraded by commodification, a virtually inescapable result of allowing baby-making to proceed under the banner of commerce. Genetic and reproductive biotechnology companies are already growth industries, but they will go into commercial orbit once the Human Genome Project nears completion. Supply will create enormous demand. Even before the capacity for human cloning arrives, established companies will have invested in the harvesting of eggs from ovaries obtained at autopsy or through ovarian surgery, practiced embryonic genetic alteration, and initiated the stockpiling of prospective donor tissues. Through the rental of surrogate-womb services, and through the buying and selling of tissues and embryos, priced according to the merit of the donor, the commodification of nascent human life will be unstoppable.

Finally, and perhaps most important, the practice of human cloning by nuclear transfer—like other anticipated forms of genetic engineering of the next generation—would enshrine and aggravate a profound and mischievous misunderstanding of the meaning of having children and of the parent-child relationship. When a couple now chooses to procreate, the partners are saying yes to the emergence of new life in its novelty, saying yes not only to having a child but also, tacitly, to having whatever child this child turns out to be. In accepting our finitude and opening ourselves to our replacement, we are tacitly confessing the limits of our control. In this ubiquitous way of nature, embracing the future by procreating means precisely that we are relinquishing our grip, in the very activity of taking up our own share in what we hope will be the immortality of human life and the human species. This means that our children are not *our* children: they are not our property, not our possessions. Neither are they supposed to live our lives for us, or anyone else's life but their own. To be sure, we seek to guide them on their way, imparting to them not just life but nurturing, love, and a way of life; to be sure, they bear our hopes that they will live fine and flourishing lives, enabling us in small measure to transcend our own limitations. Still, their genetic distinctiveness and independence are the natural foreshadowing of the deep truth that they have their own and never-before-enacted life to live. They are sprung from a past, but they take an uncharted course into the future.

Much harm is already done by parents who try to live vicariously through their children. Children are sometimes compelled to fulfill the broken dreams of unhappy parents: John Doe Jr. or the III is under the burden of having to live up to his forebear's name. Still, if most parents have hopes for their children, cloning parents will have expectations. In cloning, such overbearing parents take at the start a decisive step which contradicts the entire meaning of the open and forward-looking nature of parent-child relations. The child is given a genotype that has already lived, with full expectation that this blueprint of a past life ought to be controlling of the life that is to come. Cloning is inherently despotic, for it seeks to make one's children (or someone else's children) after one's own image (or an image of one's choosing) and their future according to one's will. In some cases, the despotism may be mild and benevolent. In other cases, it will be mischievous and downright tyrannical. But despotism—the control of another through one's will—it inevitably will be.

Meeting Some Objections

The defenders of cloning, of course, are not wittingly friends of despotism. Indeed, they regard themselves mainly as friends of freedom: the freedom of individuals to reproduce, the freedom of scientists and inventors to discover and devise and to foster "progress" in genetic knowledge and technique. They want large-scale cloning only for animals, but they wish to preserve cloning as a human option for exercising our "right to reproduce"—our right to have children, and children with "desirable genes." As law professor John Robertson points out, under our "right to reproduce" we already practice early forms of unnatural, artificial and extramarital reproduction, and we already practice early forms of eugenic choice. For this reason, he argues, cloning is no big deal.

We have here a perfect example of the logic of the slippery slope, and the slippery way in which it already works in this area. Only a few years ago, slippery slope arguments were used to oppose artificial insemination and in vitro fertilization using unrelated sperm donors. Principles used to justify these practices, it was said, will be used to justify more artificial and more eugenic practices, including cloning. Not so, the defenders retorted, since we can make the necessary distinctions. And now, without even a gesture at making the necessary distinctions, the continuity of practice is held by itself to be justificatory.

The principle of reproductive freedom as currently enunciated by the proponents of cloning logically embraces the ethical acceptability of sliding down the entire rest of the slope—to producing children ectogenetically from sperm to term (should it become feasible) and to producing children whose entire genetic makeup will be the product of parental eugenic planning and choice. If reproductive freedom means the right to have a child of one's own choosing, by whatever means, it knows and accepts no limits.

But, far from being legitimated by a "right to reproduce," the emergence of techniques of assisted reproduction and genetic engineering should compel us to reconsider the meaning and limits of such a putative right. In truth, a "right to reproduce" has always been a peculiar and problematic notion. Rights generally belong to individuals, but this is a right which (before cloning) no one can exercise alone. Does the right then inhere only in couples? Only in married couples? Is it a (woman's) right to carry or deliver or a right (of one or more parents) to nurture and rear? Is it a right to have your own biological child? Is it a right only to attempt reproduction, or a right also to succeed? Is it a right to acquire the baby of one's choice?

The assertion of a negative "right to reproduce" certainly makes sense when it claims protection against state interference with procreative liberty, say, through a program of compulsory sterilization. But surely it cannot be the basis of a tort claim against nature, to be made good by technology, should free efforts at natural procreation fail. Some insist that the right to reproduce embraces also the right against state interference with the free use of all technological means to obtain a child. Yet such a position cannot be sustained: for reasons having to do with the means employed, any community may rightfully prohibit surrogate pregnancy, or polygamy, or the sale of babies to infertile couples, without violating anyone's basic human "right to reproduce." When the exercise of a previously innocuous freedom now involves or impinges on troublesome practices that the original freedom never was intended to reach, the general presumption of liberty needs to be reconsidered.

We do indeed already practice negative eugenic selection, through genetic screening and prenatal diagnosis. Yet our practices are governed by a norm of health. We seek to prevent the birth of children who suffer from known (serious) genetic diseases. When and if gene therapy becomes possible, such diseases could then be treated, in utero or even before implantation—I have no ethical objection in principle to such a practice (though I have some practical worries), precisely because it serves the medical goal of healing existing individuals. But therapy, to be therapy, implies not only an existing "patient." It also implies a norm of health. In this respect, even germline gene "therapy," though practiced not on a human being but on egg and sperm, is less radical than cloning, which is in no way therapeutic. But once one blurs the distinction between health promotion and genetic enhancement, between so-called negative and positive eugenics, one opens the door to all future eugenic designs. "To make sure that a child will be healthy and have good chances in life": this is Robertson's principle, and owing to its latter clause it is an utterly elastic principle, with no boundaries. Being over eight feet tall will likely produce some very good chances in life, and so will having the looks of Marilyn Monroe, and so will a genius-level intelligence.

Proponents want us to believe that there are legitimate uses of cloning that can be distinguished from illegitimate uses, but by their own principles no such limits can be found. (Nor could any such limits be enforced in practice.) Reproductive freedom, as they understand it, is governed solely by the subjective wishes of the parents-to-be (plus the avoidance of bodily harm to the child). The sentimentally appealing case of the childless married couple is, on these grounds, indistinguishable from the case of an individual (married or not) who would like to clone someone famous or talented, living or dead. Further, the principle here endorsed justifies not only cloning but, indeed, all future artificial attempts to create (manufacture) "perfect" babies.

A concrete example will show how, in practice no less than in principle, the so-called innocent case will merge with, or even turn into, the more troubling ones. In practice, the eager parents-to-be will necessarily be subject to the tyranny of expertise. Consider an infertile married couple, she lacking eggs or he lacking sperm, that wants a child of their (genetic) own, and propose to clone either husband or wife. The scientist-physician (who is also co-owner of the cloning company) points out the likely difficulties—a cloned child is not really their (genetic) child, but the child of only *one* of them; this imbalance may produce strains on the marriage; the child might suffer identity confusion; there is a risk of perpetuating the cause of sterility; and so on—and he also points out the advantages of choosing a donor nucleus. Far better than a child of their own would be a child of their own choosing. Touting his own expertise in selecting healthy and talented donors, the doctor presents the couple with his latest catalog containing the pictures, the health records and the accomplishments of his stable of cloning donors, samples of whose tissues are in his deep freeze. Why not, dearly beloved, a more perfect baby?

The "perfect baby," of course, is the project not of the infertility doctors, but of the eugenic scientists and their supporters. For them, the paramount right is not the so-called right to reproduce but what biologist Bentley Glass called, a quarter of a century ago, "the right of every child to be born with a sound physical and mental constitution, based on a sound genotype . . . the inalienable right to a sound heritage." But to secure this right, and to achieve the requisite quality control over new human life, human conception and gestation will need to be brought fully into the bright light of the laboratory, beneath which it can be fertilized, nourished, pruned, weeded, watched, inspected, prodded, pinched, cajoled, injected, tested, rated, graded, approved, stamped, wrapped, sealed and delivered. There is no other way to produce the perfect baby.

Yet we are urged by proponents of cloning to forget about the science fiction scenarios of laboratory manufacture and multiple-copied clones, and to focus only on the homely cases of infertile couples exercising their reproductive rights. But why, if the single cases are so innocent, should multiplying their performance be so off-putting? (Similarly, why do others object to people making money off this practice, if the practice itself is perfectly acceptable?) When we follow the sound ethical principle of universalizing our choice—"would it be right if everyone cloned a Wilt Chamberlain (with his consent, of course)? Would it be right if everyone decided to practice asexual reproduction?"—we discover what is wrong with these seemingly innocent cases. The so-called science fiction cases make vivid the meaning of what looks to us, mistakenly, to be benign.

Though I recognize certain continuities between cloning and, say, in vitro fertilization, I believe that cloning differs in essential and important ways. Yet those who disagree should be reminded that the "continuity" argument cuts both ways. Sometimes we establish bad precedents, and discover that they were bad only when we follow their inexorable logic to places we never meant to go. Can the defenders of cloning show us today how, on their principles, we will be able to see producing babies ("perfect babies") entirely in the laboratory or exercising full control over their genotypes (including so-called, enhancement) as ethically different, in any essential way, from present forms of assisted reproduction? Or are they willing to admit, despite their attachment to the principle of continuity, that the complete obliteration of "mother" or "father," the complete depersonalization of procreation, the complete manufacture of human beings and the complete genetic control of one generation over the next would be ethically problematic and essentially different from current forms of assisted reproduction? If so, where and how will they draw the line, and why? I draw it at cloning, for all the reasons given.

Ban the Cloning of Humans

What, then, should we do? We should declare that human cloning is unethical in itself and dangerous in its likely consequences. In so doing, we shall have the backing of the overwhelming majority of our fellow Americans, and of the human race, and (I believe) of most practicing scientists. Next, we should do all that we can to prevent the cloning of human beings. We should do this by means of an international legal ban if possible, and by a unilateral national ban, at a minimum. Scientists may secretly undertake to violate such a law, but they will be deterred by not being able to stand up proudly to claim the credit for their technological bravado and success. Such a ban on clonal baby-making, moreover, will not harm the progress of basic genetic science and technology. On the contrary, it will reassure the public that scientists are happy to proceed without violating the deep ethical norms and intuitions of the human community.

This still leaves the vexed question about laboratory research using early, embryonic human clones, specially created only for such research purposes, with no intention to implant them into a uterus. There is no question that such research holds great promise for gaining fundamental knowledge about normal (and abnormal) differentiation, and for developing tissue lines for transplantation that might be used, say, in treating leukemia or in repairing brain or spinal cord injuries—to mention just a few of the conceivable benefits. Still, unrestricted clonal embryo research will surely make the production of living human clones much more likely. Once the genies put the cloned embryos into the bottles, who can strictly control where they go (especially in the absence of legal prohibitions against implanting them to produce a child)?

I appreciate the potentially great gains in scientific knowledge and medical treatment available from embryo research, especially with cloned embryos. At the same time, I have serious reservations about creating human embryos for the sole purpose of experimentation. There is something deeply repugnant and fundamentally transgressive about such a utilitarian treatment of prospective human life. This total, shameless exploitation is worse, in my opinion, than the "mere" destruction of nascent life. But I see added objections, as a matter of principle, to creating and using *cloned* early embryos for research purposes, beyond the objections that I might raise to doing so with embryos produced sexually.

And yet, as a matter of policy and prudence, any opponent of the manufacture of cloned humans must, I think, in the end oppose also the creating of cloned human embryos. Frozen embryonic clones (belonging to whom?) can be shuttled around without detection. Commercial ventures in human cloning will be developed without adequate oversight. In order to build a fence around the law, prudence dictates that one oppose—for this reason alone—all production of cloned human embryos, even for research purposes. We should allow all cloning research on animals to go forward, but the only safe trench that we can dig across the slippery slope, I suspect, is to insist on the inviolable distinction between animal and human cloning.

Some readers, and certainly most scientists, will not accept such prudent restraints, since they desire the benefits of research. They will prefer, even in fear and trembling, to allow human embryo cloning research to go forward.

Very well. Let us test them. If the scientists want to be taken seriously on ethical grounds, they must at the very least agree that embryonic research may proceed if and only if it is preceded by an absolute and effective ban on all attempts to implant into a uterus a cloned human embryo (cloned from an adult) to produce a living child. Absolutely no permission for the former without the latter.

The National Bioethics Advisory Commission's recommendations regarding this matter should be watched with the greatest care. Yielding to the wishes of the scientists, the commission will almost surely recommend that cloning human embryos for research be permitted. To allay public concern, it will likely also call for a temporary moratorium—not a legislative ban—on implanting cloned embryos to make a child, at least until such time as cloning techniques will have been perfected and rendered "safe" (precisely through the permitted research with cloned embryos). But the call for a moratorium rather than a legal ban would be a moral and a practical failure. Morally, this ethics commission would (at best) be waffling on the main ethical question, by refusing to declare the production of human clones unethical (or ethical). Practically, a moratorium on implantation cannot provide even the minimum protection needed to prevent the production of cloned humans.

Opponents of cloning need therefore to be vigilant. Indeed, no one should be willing even to consider a recommendation to allow the embryo research to proceed unless it is accompanied by a call for *prohibiting* implantation and until steps are taken to make such a prohibition effective.

Technically, the National Bioethics Advisory Commission can advise the president only on federal policy, especially federal funding policy. But given the seriousness of the matter at hand, and the grave public concern that goes beyond federal funding, the commission should take a broader view. (If it doesn't, Congress surely will.) Given that most assisted reproduction occurs in the private sector, it would be cowardly and insufficient for the commission to say, simply, "no federal funding" for such practices. It would be disingenuous to argue that we should allow federal funding so that we would then be able to regulate the practice; the private sector will not be bound by such regulations. Far better, for virtually everyone concerned, would be to distinguish between research on embryos and baby-making, and to call for a complete national and international ban (effected by legislation and treaty) of the latter, while allowing the former to proceed (at least in private laboratories).

The proposal for such a legislative ban is without American precedent, at least in technological matters, though the British and others have banned cloning of human beings, and we ourselves ban incest, polygamy and other forms of "reproductive freedom." Needless to say, working out the details of such a ban, especially a global one, would be tricky, what with the need to develop appropriate sanctions for violators. Perhaps such a ban will prove ineffective; perhaps it will eventually be shown to have been a mistake. But it would at least place the burden of practical proof where it belongs: on the proponents of this horror, requiring them to show very clearly what great social or medical good can be had only by the cloning of human beings.

We Americans have lived by, and prospered under, a rosy optimism about scientific and technological progress. The technological imperative—if it can be done, it must be done—has probably served us well, though we should admit that there is no accurate method for weighing benefits and harms. Even when, as in the cases of environmental pollution, urban decay or the lingering deaths that are the unintended by-products of medical success, we recognize the unwelcome outcomes of technological advance, we remain confident in our ability to fix all the "bad" consequences—usually by means of still newer and better technologies. How successful we can continue to be in such post hoc repairing is at least an open question. But there is very good reason for shifting the paradigm around, at least regarding those technological interventions into the human body and mind that will surely effect fundamental (and likely irreversible) changes in human nature, basic human relationships, and what it means to be a human being. Here we surely should not be willing to risk everything in the naive hope that, should things go wrong, we can later set them right.

The president's call for a moratorium on human cloning has given us an important opportunity. In a truly unprecedented way, we can strike a blow for the human control of the technological project, for wisdom, prudence and human dignity. The prospect of human cloning, so repulsive to contemplate, is the occasion for deciding whether we shall be slaves of unregulated progress, and ultimately its artifacts, or whether we shall remain free human beings who guide our technique toward the enhancement of human dignity. If we are to seize the occasion, we must, as the late Paul Ramsey wrote,

> raise the ethical questions with a serious and not a frivolous conscience. A man of frivolous conscience announces that there are ethical

quandaries ahead that we must urgently consider before the future catches up with us. By this he often means that we need to devise a new ethics that will provide the rationalization for doing in the future what men are bound to do because of new actions and interventions science will have made possible. In contrast a man of serious conscience means to say in raising urgent ethical questions that there may be some things that men should never do. The good things that men do can be made complete only by the things they refuse to do.

Part II
Perspectives from Religion

Perspectives from Religion: Introduction

THE MOST VOCAL critics of human cloning have been representatives of religious positions, whose arguments challenge cherished assumptions, such as the notion that human cloning is simply a private matter without profound social consequences, as well as our views of the totality of the human person. Religious perspectives pose tough questions: What is the moral status of the pre-embryo? Is there a natural link between sexual intercourse and procreation? Does the ultimately narcissistic character of cloning point to a real difference between begetting and making, and, if so, how does this affect our views of the child? Issues of justice also have theological bearings. Minority groups, especially African Americans, have been exploited under the banner of medical progress. With human cloning, is there any guarantee that such abuse will not continue?

Religious perspectives remain prominent in our culture, a culture which also thrives on a plurality of religious beliefs. So, according to the Jewish tradition, what is the moral and legal status of the human clone? A Native American perspective seems to consider cloning within a broader scope by inquiring into its far-reaching, practical consequences. The Orthodox Christian viewpoint stresses the sacramental nature of marriage, and asks, Does human cloning violate this sacramental character? Is human cloning compatible with Muslim teachings? How does human cloning, as a novel form of reproduction, relate to Buddhist views of rebirth? Is it at all consistent with the Hindu four aims in life and its central ethical teaching of karma?

And although many other arguments against human cloning are secular in expression, are they genuinely secular, or do they disguise a more fundamentally religious disposition? One of the key theological arguments holds that human cloning is a violation of what is God-given and natural. Yet what is meant by natural? In this vein, do we tend to give more authority than we should to religious arguments *because* they are religious?

Richard McCormick writes in the wake of the cloning of human embryos at The George Washington University Medical Center on October 1993, and he specifically attacks arguments in support of human cloning that appeal to individual rights and autonomy. To assume automatically that any means to overcome infertility is morally legitimate misses the point, he argues, because human cloning is not merely a private matter. It is significantly social, and McCormick singles out some dimensions that underscore this sociality.

To begin with, as we gain more control over the selection of our offspring, we stand in danger of viewing them primarily in terms of specific traits such as intelligence, looks, strength, height, and eye color. This preferential breeding is demeaning to their totality as whole persons. Along these lines, cloning is a giant step toward reducing the worth of human beings to mere genetic material, thus threatening their individuality. Moreover, McCormick argues, because the pre-embryo possesses the potential for personhood and because there is general uncertainty as to the precise nature of its status, it should be considered a person and treated as such. This judgment requires that we be extremely circumspect regarding research on pre-embryos, and McCormick advocates a uniform national mechanism for monitoring such research.

Gilbert Meilaender weaves his case against human cloning from a variety of theological angles. A critical starting point is that a child, equal to all other human persons, is a gift and not a product. He argues that there is enough evidence in Genesis 1 to establish the primacy of the link between sexual differentiation and procreation; cloning strikes at the core of this link and splits it apart. The normative position among Protestants is that the sexual union is oriented toward procreation, which should occur within the legitimate context of marriage, and the child is therefore a gift resulting from this sexual union. The normative link is thus triadic: sexual differentiation–marriage–procreation.

Why is this link normative? According to Meilaender, it authenticates the couple's relationship to each other. If their sexual union is a natural expression of their love, it transcends their own "personal project," in a sense expressing a sort of third person—their love—embodied in the form of the child. The operative word here is love, for the act of genuinely loving another necessitates the transcendence of self-interests. In this respect, the child born out of love is not a project.

The crucial distinction is that between begetting and making. Meilaender asserts that begetting presumes the child's equality with its parents, whereas making presumes an inequality in that the child has become the project of the parents, a product and instrument of their will. Having a child that is the couple's project is narcissistic. And although Meilaender admits the possible benefits of experimentation upon human embryos, he still claims, as does McCormick, that since we are not certain about the moral status of this "unimplanted embryo," it makes sense to prohibit any further research.

Marian Gray Secundy speaks out on behalf of ethnic Americans, particularly African Americans who have numerous reasons to distrust medical progress and research in America because of their sad history of being the unwilling subjects of experimentation. The new technologies pertaining to human cloning are equally capable of being abused in ways that would exploit vulnerable groups.

While she therefore opts to ban human cloning, Secundy also proposes to educate and inform the general public about cloning and its consequences. And this program includes an all-out effort at grassroots education, with special attention to the concerns of ethnic Americans.

Kenneth Robinson welcomes medical advances that would improve the human lot by preventing genetic diseases such as sickle-cell anemia. At the same time, like Secundy, he reminds us that the connotations of eugenics evoke apprehension, particularly for vulnerable groups such as African Americans. Although African Americans welcomed ways to circumvent infertility, for instance, many of them were deprived of fair access to the technologies. Furthermore, because many African Americans bestow a high moral status upon human embryos, research conducted on human embryos becomes problematic.

Given the dominant societal ethos of commercialism, as well as sustained expressions of racism, Robinson argues for strict regulatory monitoring of any cloning research. Otherwise, vulnerable groups are the ones most at risk of potential abuses such as inequitable access and unethical research.

Speaking on behalf of Orthodox Christianity, Father Stanley Harakas takes a strong stand against human cloning, charging that it involves the blatant manufacturing and exploitation of a human being. As a product solely of human design and laboratory procedure, the cloned human being would be technically deprived of parents in the genuine sense. Furthermore, the laboratory method of reproduction destroys the sacramental quality of marriage.

According to Harakas, there is no doubt that a clone would have a soul. Yet research continued to the point of combining nonhuman elements (for example, nonhuman DNA) with human DNA would further desacralize the dignity of the human being. The precedent for this mixture of elements has already occurred with the Loma Linda University xenografts on an infant girl. For Harakas, such combination oversteps an inviolable boundary and cannot be justified.

What insights regarding human cloning can be gathered from Jewish teachings? According to Rabbi Barry Freundel, one of the more significant issues concerns the legal and moral status of the clone. Employing the analogy of the Golem tales in Jewish mystical tradition, he argues that there is every reason to assign personal and genuine moral status to the human clone. Yet because parental responsibilities become more confusing once we sever the traditional link between procreation and marriage, the issue of assigning parental responsibilities becomes problematic.

Freundel is convinced that there must be safeguards against potential abuses, but he demonstrates that Judaic teachings evidence a more optimistic attitude toward cloning, as long as strict regulatory guidelines are in place.

As for Islamic teachings, Maher Hathout recognizes that quite a few Muslim scholars endorse a ban on human cloning, a position akin to that of the Vatican. He contends, however, that research is part of Allah's design and should not be prohibited. His position is grounded in the Qur'an's teachings, which stipulate that research and human investigation are consistent with Allah's will. Allah has given humans the capacity to create, and science is in accord with Allah's will so long as it improves the human condition. Research on human cloning, therefore, is not in opposition to the will of God, as most religious commentators seem to believe.

Nevertheless, Hathout offers one critical caveat, warning against the potential for commercialization of cloning research. Islamic teachings provide strict prohibitions against such abuse of scientific investigation, and Hathout's warning is apropos in view of today's culture of commercialism.

Other religious voices heard in this section include Viola Cordova, a Native American who challenges two assumptions: that individual rights and interests take precedence over the common good, and that knowledge in itself is always good. She stresses that any actions we do, either as individuals or as a group, create ripple effects like those caused by a pebble thrown into a pond. Her analogy is extremely pragmatic, and it compels us to reconsider our intended actions in light of the consequences they may have upon the welfare of the group as well as on other aspects of the environment.

Cordova offers another highly practical point when she argues that, in the context of our already overpopulated planet—including the acutely increasing population in the United States—efforts at human cloning can only compound the demographic problem and are thereby counterproductive.

Can Buddhist teachings shed further light on the morality of human cloning? Ronald Nakasone tells us that, in the light of traditional Buddhist teachings, human cloning presents an interesting challenge. For instance, he contrasts human cloning with Vasubandhu's teaching in the revered text, the *Abhidharmakosa*, which explains Buddhist views of reproduction and rebirth in relation to an intermediate state. At the same time, Buddhists teach that the nature of reality consists of change, which prompts adaptation of moral evaluations within the framework of novel conditions brought about by medical and technological discoveries. An example is the altered demographic dynamic created by the increasing global population of elderly. This perspective requires us to view human cloning in broader contexts.

Finally, what about Hindu teachings? Arvind Sharma emphasizes the more tolerant nature of Hinduism's religious and philosophical system, which is open to a plurality of viewpoints and interpretations on various issues. Human cloning can be understood within what Hindus believe are the four primary goals in life: living morally, maintaining an appropriate level of material comfort, experiencing sexual well-being, and cultivating spiritual enlightenment. All of these goals are regulated by *dharma*, or morality—which is where karma, the principle of moral cause-and-effect, comes into play. The key premise in karma is that each individual is unique and thereby accountable to him- or herself. For this reason, children who result from human cloning would also be unique and subject to the same karmic laws. Furthermore, the Hindu belief that there is no strict distance between the so-called supernatural and the natural in the creative process influences views on human cloning.

As a scientist, Lee Silver explores some of the typical secular arguments against human cloning, such as the likelihood of physical harm to the clone. This assumption, he argues, is unfounded. As to the concern about possible psychological harm due to the threat cloning poses to the clone's individuality and identity, he points out that the experience of natural identical twins shows this to be fallacious.

Silver maintains that although much of the opposition to human cloning is posed in secular guise, it actually runs deeper. He directs our attention to what he feels is the *real* reason: the religious notion that cloning is a case of humans acting out of hubris, and attempting to be more like God. Resistance to cloning is thus actually a negative reaction against our Promethean pride. Silver then argues that this deep-rooted objection assumes a particular view of God that is not at all universal, and he contends that it would be foolish to base policy upon narrow religious premises.

In the same spirit as Silver, Richard Dawkins, another respected scientist, deftly counters the typical arguments against human cloning. In the process, he challenges our notions regarding what is natural and contends that the *onus probandum,* or burden of proof, lies with those who oppose human cloning as being either unnatural or causing more harm than good. The burden of proof is usually viewed the other way around, in part reflecting the special privilege we assign to religious arguments. He shares Silver's view that the real bottom line of contention against cloning stems from religious objections, but he goes one step further, in that he decries the tendency to accord a special, nearly infallible status to positions based upon religious beliefs and premises. And Dawkins minces no words when he chastises religious lobbies as well as those who claim to be spokespersons for religious positions in opposition to cloning yet who lack sufficient knowledge about the practice.

Should We Clone Humans?

by Richard A. McCormick, S.J.

Richard A. McCormick, S.J. was a Professor of Theology at the University of Notre Dame. He recently passed away.

THE CLONING OF human embryos by Dr. Jerry L. Hall at George Washington University Medical Center last month has set off an interesting ethical debate. Should it be done? For what purposes? With what controls? It is not surprising—though I find it appalling—that some commentators see the entire issue in terms of individual autonomy. Embryos belong to their producers, they argue, and it is not society's business to interfere with the exercise of people's privacy (see comments in the *New York Times*, October 26).

One's approach to cloning will vary according to the range of issues one wants to consider. For example, some people will focus solely on the role of cloning in aiding infertile couples—and they will likely conclude that there is nothing wrong with it. The scarcely hidden assumption is that anything that helps overcome infertility is morally appropriate. That is, I believe, frighteningly myopic.

Human cloning is an extremely social matter, not a question of mere personal privacy. I see three dimensions to the moral question: the wholeness of life, the individuality of life, and respect of life.

Wholeness

Our society has gone a long way down the road of positive eugenics, the preferential breeding of superior genotypes. People offhandedly refer to "the right to a healthy child." Implied in such loose talk is the right to discard the imperfect. What is meant, of course, is that couples have a claim to reasonably available means to ensure that their children are born healthy. We have preimplementation diagnosis for genetic defects. We have recently seen several cases of "wrongful life" where the child herself or himself is the plaintiff. As a member of the ethics committee of the American Fertility Society, I regularly receive brochures from sperm banks stating the donor's race, education, hobbies, height, weight and eye color. We are rapidly becoming a pick-and-choose society with regard to our prospective children. More than a few couples withhold acceptance of their fetuses pending further testing.

This practice of eugenics raises a host of problems: What qualities are to be maximized? What defects are intolerable? Who decides? But the critical flaw in "preferential breeding" is the perversion of our attitudes: we begin to value the person in terms of the particular trait he or she was programmed to have. In short, we reduce the whole to a part. People who do that are in a moral wilderness.

Individuality

Uniqueness and diversity (sexual, racial, ethnic, cultural) are treasured aspects of the human condition, as was sharply noted by a study group of the National Council of Churches in 1984 (*Genetic Engineering: Social and Ethical Consequences*). Viewed theologically, human beings, in their enchanting, irreplaceable uniqueness and with all their differences, are made in the image of God. Eugenics schemes that would bypass, downplay or flatten human diversities and uniqueness should be viewed with a beady eye. In the age of the Genome Project it is increasingly possible to collapse the human person into genetic data. Such reductionism could shatter our wonder at human individuality and diversity at the very time that, in other spheres of life, we are emphasizing it.

Life

Everyone admits that the pre-embryo (preimplanted embryo) is human life. It is living, not dead. It is human, not canine. One need not attribute personhood to such early life to claim that it demands respect and protection.

Two considerations must be carefully weighed as we try to discern our obligations toward pre-embryonic life. The first consideration is for the potential of the pre-embryo. Under favorable circumstances, the fertilized ovum will move through developmental individuality and then progressively through functional, behavioral, psychic and social individuality. In viewing the first stage, one cannot afford to blot out subsequent stages. It retains its potential for personhood and thus deserves profound respect. This is a weighty matter for the believer who sees the human person as a member of God's family and the temple of the Spirit. Interference with such a potential future cannot be a light undertaking.

The second consideration concerns our own human condition. I would gather these concerns under the notion of "uncertainty." There is uncertainty about the extent to which enthusiasm for human research can be controlled. That is, if we concluded that pre-embryos need not be treated in all circumstances as persons, would we little by little extend this to embryos? Would we gradually trivialize the reasons justifying pre-embryo manipulation? These are not abstract worries; they have become live questions.

Furthermore, there is uncertainty about the effect of pre-embryo manipulation on personal and societal attitudes toward nascent human life in general. Will there be further erosion of our respect? I say "further" because of the widespread acceptance and practice of abortion. There is grave uncertainty about our ability to say no and backtrack when we detect abuses, especially if they have produced valuable scientific and therapeutic data or significant treatment. Medical technology ("progress") has a way of establishing irreversible dynamics.

Because the pre-embryo does have intrinsic potential and because of the many uncertainties noted above, I would argue that the pre-embryo should be treated as a person. These obligations may be prima facie—to use W.D. Ross's phrase—and subject to qualifications. But when we are dealing with human life, the matter is too important to be left to local or regional criteria and controls. We need uniform national controls. Without them, our corporate reverence for life, already so deeply compromised, will be further eroded.

In sum, human cloning is not just another technological step to be judged in terms of its effects on those cloned. What frightens me above all is what human cloning would do to all of us—to our sense of the wholeness, individuality, and sanctity of human life. These are intertwined theological concerns of the first magnitude.

Begetting and Cloning

by Gilbert Meilaender

Gilbert Meilaender, a former dean of Oberlin College, now holds the Board of Directors Chair in Theological Ethics at Valparaiso University. He was invited to present his views on human cloning to the NBAC.

The following remarks were presented to the National Bioethics Advisory Commission on March 13, 1997.

I HAVE BEEN invited, as I understand it, to speak today specifically as a Protestant theologian. I have tried to take that charge seriously, and I have chosen my concerns accordingly. I do not suppose, therefore, that the issues I address are the only issues to which you ought to give your attention. Thus, for example, I will not address the question of whether we could rightly conduct the first experiments in human cloning, given the likelihood that such experiments would not at first fully succeed. That is an important moral question, but I will not take it up. Nor do I suppose that I can represent Protestants generally. There is no such beast. Indeed, Protestants are specialists in the art of fragmentation. In my own tradition, which is Lutheran, we commonly understand ourselves as quite content to be Catholic except when, on certain questions, we are compelled to disagree. Other Protestants might think of themselves differently.

More important, however, is this point: Attempting to take my charge seri-

Reprinted from *First Things*, vol. 74 (June–July 1997): 41–43.

ously, I will speak theologically—not just in the standard language of bioethics or public policy. I do not think of this, however, simply as an opportunity for the "Protestant interest group" to weigh in at your deliberations. On the contrary, this theological language has sought to uncover what is universal and human. It begins epistemologically from a particular place, but it opens up ontologically a vision of the human. The unease about human cloning that I will express is widely shared. I aim to get at some of the theological underpinnings of that unease in language that may seem unfamiliar or even unwelcome, but it is language that is grounded in important Christian affirmations that seek to understand the child as our equal—one who is a gift and not a product. In any case, I will do you the honor of assuming that you are interested in hearing what those who speak such a language have to say, and I will also suppose that a faith which seeks understanding may sometimes find it.

Lacking an accepted teaching office within the church, Protestants had to find some way to provide authoritative moral guidance. They turned from the authority of the church as interpreter of Scripture to the biblical texts themselves. That characteristic Protestant move is not likely, of course, to provide any very immediate guidance on a subject such as human cloning. But it does teach something about the connection of marriage and parenthood. The creation story in the first chapter of Genesis depicts the creation of humankind as male and female, sexually differentiated and enjoined by God's grace to sustain human life through procreation.

Hence, there is given in creation a connection between the differentiation of the sexes and the begetting of a child. We begin with that connection, making our way indirectly toward the subject of cloning. It is from the vantage point of this connection that our theological tradition has addressed two questions that are both profound and mysterious in their simplicity: What is the meaning of a child? And what is good for a child? These questions are, as you know, at the heart of many problems in our society today, and it is against the background of such questions that I want to reflect upon the significance of human cloning. What Protestants found in the Bible was a normative view: namely, that the sexual differentiation is ordered toward the creation of offspring, and children should be conceived within the marital union. By God's grace the child is a gift who springs from the giving and receiving of love. Marriage and parenthood are connected—held together in a basic form of humanity.

To this depiction of the connection between sexual differentiation and child-bearing as normative, it is, as Anglican theologian Oliver O'Donovan has argued, possible to respond in different ways. We may welcome the connection and find in it humane wisdom to guide our conduct. We may resent it as a limit to our freedom and seek to transcend it. We did not need modern scientific breakthroughs to know that it is possible—and sometimes seemingly desirable—to sever the connection between marriage and begetting children. The possibility of human cloning is striking only because it breaks the connection so emphatically. It aims directly at the heart of the mystery that is a child. Part of the mystery here is that we will always be hard-pressed to explain why the connection of sexual differentiation and procreation should not be broken. Precisely to the degree that it is a basic form of humanity, it will be hard to give more fundamental reasons why the connection should be welcomed and honored when, in our freedom, we need not do so. But moral argument must begin somewhere. To see through everything is, as C.S. Lewis once put it, the same as not to see at all.

If we cannot argue to this starting point, however, we can argue from it. If we cannot entirely explain the mystery, we can explicate it. And the explication comes from two angles. Maintaining the connection between procreation and the sexual relationship of a man and woman is good both for that relationship and for children.

It is good, first, for the relation of the man and woman. No doubt the motives of those who beget children coitally are often mixed, and they may be uncertain about the full significance of what they do. But if they are willing to shape their intentions in accord with the norm I have outlined, they may be freed from self-absorption. The act of love is not simply a personal project undertaken to satisfy one's own needs, and procreation, as the fruit of coitus, reminds us of that. Even when the relation of a man and woman does not or cannot give rise to offspring, they can understand their embrace as more than their personal project in the world, as their participation in a form of life that carries its own inner meaning and has its telos established in the creation. The meaning of what we do then is not determined simply by our desire or will. As Oliver O'Donovan has noted, some understanding like this is needed if the sexual relation of a man and woman is to be more than "simply a profound form of play."

And when the sexual act becomes only a personal project, so does the child. No longer then is the bearing and rearing of children thought of as a task we should take up or as a return we make for the gift of life; instead, it is a project we undertake if it promises to meet our needs and desires. Those people—both learned commentators and ordinary folk—who in recent days have described cloning as narcissistic or as replication of one's self see something important. Even if we grant that a clone, reared in different circumstances than its immediate ancestor, might turn out to be quite a different person in some respects, the point of that person's existence would be grounded in our will and desire.

Hence, retaining the tie that unites procreation with the sexual relation of a man and woman is also good for children. Even when a man and woman deeply desire a child, the act of love itself cannot take the child as its primary object. They must give themselves to each other, setting aside their projects, and the child becomes the natural fruition of their shared love—something quite different from a chosen project. The child is therefore always a gift—one like them who springs from their embrace, not a being whom they have made and whose destiny they should determine. This is light years away from the notion that we all have a right to have children—in whatever way we see fit, whenever it serves our purposes. Our children begin with a kind of genetic independence of us, their parents. They replicate neither their father nor their mother. That is a reminder of the independence that we must eventually grant to them and for which it is our duty to prepare them. To lose, even in principle, this sense of the child as a gift entrusted to us will not be good for children.

I will press this point still further by making one more theological move. When Christians tried to tell the story of Jesus as they found it in their Scriptures, they were driven to some rather complex formulations. They wanted to say that Jesus was truly one with that God whom he called Father, lest it should seem that what he had accomplished did not really overcome the gulf that separates us from God. Thus, while distinguishing the persons of Father and Son, they wanted to say that Jesus is truly God—of one being with the Father. And the language in which they did this (in the fourth-century Nicene Creed, one of the two most important creeds that antedate the division of the church in the West at the Reformation) is language which describes the Son of the Father as "begotten, not made." Oliver O'Donovan has noted that this distinction between making and begetting, crucial for Christians' understanding of God, carries considerable moral significance.

What the language of the Nicene Creed wanted to say was that the Son is God just as the Father is God. It was intended to assert an equality of being. And for that what was needed was a language other than the language of making. What we beget is like ourselves. What we make is not; it is the product of our free decision, and its destiny is ours to determine. Of course, on this Christian understanding human beings are not begotten in the absolute sense that the Son is said to be begotten of the Father. They are made—but made by God through human begetting.

Hence, although we are not God's equal, we are of equal dignity with each other. And we are not at each other's disposal. If it is, in fact, human begetting that expresses our equal dignity, we should not lightly set it aside in a manner as decisive as cloning.

I am well aware, of course, that other advances in what we are pleased to call reproductive technology have already strained the connection between the sexual relationship of a man and woman and the birth of a child. Clearly, procreation has to some extent become reproduction, making rather than doing. I am far from thinking that all this has been done well or wisely, and sometimes we may only come to understand the nature of the road we are on when we have already traveled fairly far along it. But whatever we say of that, surely human cloning would be a new and decisive turn on this road—far more emphatically a kind of production, far less a surrender to the mystery of the genetic lottery which is the mystery of the child who replicates neither father nor mother but incarnates their union, far more an understanding of the child as a product of human will.

I am also aware that we can all imagine circumstances in which we ourselves might—were the technology available—be tempted to turn to cloning. Parents who lose a child in an accident and want to "replace" her. A seriously ill person in need of embryonic stem cells to repair damaged tissue. A person in need of organs for transplant. A person who is infertile and wants, in some sense, to reproduce. Once the child becomes a project or product, such temptations become almost irresistible. There is no end of good causes in the world, and they would sorely tempt us even if we did not live in a society for which the pursuit of health has become a god, justifying almost anything.

As theologian and bioethicist William F. May has often noted, we are preoccupied with death and the destructive powers of our world. But without in any way glorifying suffering or pretending that it is not evil, Christians worship a God who wills to be with us in our dependence, teaching us "attentiveness before a good and nurturant God." We learn therefore that what matters is how we live, not only how long—that we are responsible to do as much good as we can, but this means, as much as we can within the limits morality sets for us.

I am also aware, finally, that we might for now approve human cloning but only in restricted circumstances—as, for example, the cloning of preimplantation embryos (up to fourteen days) for experimental use. That would, of course, mean the creation solely for purposes of research of human embryos—human subjects who are not really best described as preimplantation embryos. They are unimplanted embryos—a locution that makes clear the extent to which their being and destiny are the product of human will alone. If we are genuinely baffled about how best to describe the moral status of that human subject who is the unimplanted embryo, we should not go forward in a way that peculiarly combines metaphysical bewilderment with practical certitude by approving even such limited cloning for experimental purposes.

Protestants are often pictured—erroneously in many respects—as stout defenders of human freedom. But whatever the accuracy of that depiction, they have not had in mind a freedom without limit, without even the limit that is God. They have not located the dignity of human beings in a self-modifying freedom that knows no limit and that need never respect a limit which it can, in principle, transgress. It is the meaning of the child—offspring of a man and woman, but a replication of neither; their offspring, but not their product whose meaning and destiny they might determine—that, I think, constitutes such a limit to our freedom to make and remake ourselves. In the face of that mystery I hope that your Commission will remember that "progress" is always an optional goal in which nothing of the sacred inheres.

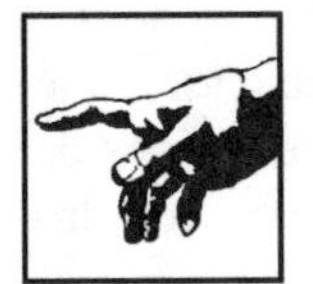

Can Science Be Trusted?

by Marian Gray Secundy

Marian Gray Secundy, former Senior Scholar at the Pritzker School of Medicine, McLean Center of Clinical Ethics, University of Chicago, now directs the Program in Clinical Ethics at Howard University's Health Sciences Center in Washington, D.C.

ETHNIC AMERICANS ARE extraordinarily suspicious and distrustful of *any* new scientific technologies. This is particularly true for, but not confined to, the African American community. The history of scientific abuse and medical neglect carries with it a legacy that is permanently imprinted upon the collective consciousness of these groups.

Inevitably, our students and residents react negatively when any issues about genetic technologies and/or related subjects are raised. The prevailing sentiment is that scientists cannot be trusted, that white scientists particularly are dangerous, that abuses are inevitable and that all manner of evil can and will most likely be visited upon the most vulnerable, e.g., ethnic groups and the poor. A family practice resident commented, "*The White man has a God complex.*" Others raised concerns about possible abuses, among them that "they" would clone soldiers for war, making "*us*" subservient tools. Of fourteen family practice residents, one-half stated that human cloning should be prohibited across the board.

Our students make allusions to religious concerns regarding human cloning and/or genetic technologies. One resident queried, "*What's left for God to do?*"

Reprinted with permission from the Program for Ethics, Science, and the Environment, Oregon State University, 1998.

Another asked, "*Where is God?*" A consensus among those with greatest concerns was that God created human beings. While we have free will, there is little doubt that in the case of cloning free will will be turned to evil and/or bad use and abused. A better option is to limit the possibility by banning cloning of humans entirely. These are the voices of East Indians, Middle Easterners, Filipinos, and Africans, as well as African Americans.

Of first order then is the requirement that we acknowledge and attend to the issues of suspicion, distrust, and fear. A guarantee of sanctionable guidelines is critical. As we have learned with issues related to organ donations, a major educational and information campaign on cloning is necessary for both consumers/patients and providers. If people are given information to help them understand what is being done, what it is possible to do, and what the implications are for now and in the future, they may become less fearful and apprehensive. Significant resources should be made available for provider and consumer education with a special focus on the unique needs and concerns of ethnic Americans. Community education and outreach should occur with the public schools, churches, civic and social groups. Dismissal of suspicion, distrust and what some might term paranoia is not appropriate or productive. Reassurance in the form of effective public policy and educational initiatives is appropriate and required.

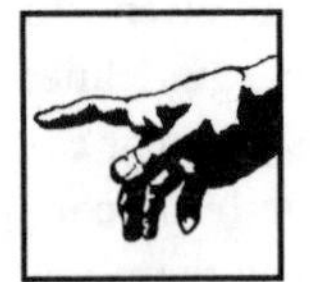

Regulating Cloning Technologies

by Rev. Kenneth S. Robinson

Rev. Kenneth S. Robinson, Pastor of St. Andrew African Methodist Episcopal Church in Memphis, Tennessee, is also Assistant Dean at the University of Tennessee College of Medicine.

AS A MINISTER/PRACTICAL theologian and a physician, I'm clear that it is inappropriate and regressive to retreat in a reflexive way to the days of the Middle Ages, to recreate the tension between institutional medicine and institutional religion; a tension which has historically appeared and reappeared even in the centuries since then. I view science and technology as tools, as methodologies which have been developed and applied—at least with the *permissive* will of God, if not always specifically and expressly with divine direction. Technology, then, becomes simply another facet of human agency—to benefit, to serve or alternatively to be a detriment to human existence.

It is the innovation *behind* the research and development of the technology, and the *application* of the technology, which create reasons for concern. The (sinful) nature of humans, and the sordid history of the interface between medicine and people of color, suggest that an *ethical universalism* does not always motivate medical researchers. It's not even the *denotation* of "eugenics" that would be of concern to African Americans; for the African American community would clearly welcome "genetic improvement" in medical conditions relatively unique to African Americans; ranging from life-threatening sickled hemo-

---

Reprinted with permission from the Program for Ethics, Science, and the Environment, Oregon State University, 1998.

globin, to our simply annoying lactase deficiency. Such genetic improvement would *benefit* African Americans. However, it is in the historical context of the *connotation* of "eugenics"—and its morally indefensible *application*—that concern arises. Surely, contemporary temptations still abound—driven by an *ethical egoism,* wrapped in *commercialism* and profit motives and/or *racism.* And, in the *current* sociopolitical context—from the emergence of a politically strong neoconservatism, to the resurgence of a significant vestige of neo-naziism—African Americans would have an understandable concern. Potential abuses of cloning technologies could only be prevented (or minimized) through

- legal/legislative controls, regulations, and *enforcement;*
- strict protocols and monitoring by the scientific community; and,
- aggressive, public oversight.

There is appreciation of the growing numbers of African American families that have overcome the challenge, the frustration and psychological pain of involuntary infertility through *in vitro* methodology, and other (low-tech) OB/GYN medical and surgical interventions. Sensitivity to the plight of infertile couples and to the potential offered by invasive or innovative procedures would lead caring, pastoral advisors to encourage the exploration of those possibilities. What *would* be of great concern would be if the technology were available to assist couples dealing with infertility, augmenting their chances of reproduction, but *access* to such was provided *selectively,* particularly to the *exclusion* of African Americans.

African Americans *tend* to be relatively conservative *vis-à-vis* biblical interpretation; i.e., allowing that life begins at "conception." Given that widely held understanding, then, the use of human embryos for medical research is problematic; particularly since *this* research involves—by definition—*living* human embryos, rather than embryonic material. Several corollary problems also evolve, including:

- the mechanism of "harvesting" such as embryos, and
- the requisite "wastage" of those embryos in the course of the research.

Given my concern about safeguarding against *exclusive* access to potential *benefits* of such research, I would not support a moratorium of federal funding for cloning research, were private research allowed to be conducted. However, my previously-stated concerns about the use of living human embryos may preempt support for effective research in *either* the public or private sector. Clearly, I believe African Americans would support federal regulatory oversight, should research continue—*regardless* of whether in the public or private sectors. I offer the following recommendations:

- the impetus and ethical motivations behind human cloning be meticulously monitored;
- the application of the research and resultant practices be tightly regulated;
- the cloning of human cells not be allowed to benefit any individual racial or ethnic or other demographic subgroup (i.e., gender), *outside of the context of a clearly identified, morally defensible, ethically sound, medically justifiable* condition which would benefit from such technology;
- access to the benefits of such technology be universal.

As advocates, spokespersons, and interpreters of God's Word for African American communities, the African American clergy—as well as the community of African American theologians—must be integrally involved in

- the formulation of an appropriate theological construct regarding cloning,
- the determination of a sound biblical hermeneutic on the issue, and
- the process of strict, reflective, and dialogical public oversight of the practice.

Human and Divine Responsibility

by Rabbi Barry Freundel

Rabbi Barry Freundel is Rabbi of the Georgetown Synagogue in Washington, D.C. He is also an Adjunct Professor of Law at Georgetown University.

I BELIEVE THERE are two main questions to human cloning. (1) Whether to proceed with cloning technology? I maintain that with appropriate safeguards, we should exercise the capacity to go ahead, while raising questions about the "upside to cloning" in terms of its scientific and human rationale. (2) More important is the question of the moral and legal status of a clone, and on this the Jewish tradition would decisively say that a clone is a human being.

The Golem stories in Jewish mystical tradition speak to the question of the status of human life without human parentage. They provide some analogical parallels with a human clone, although the analogy is not complete. The stories describe "artificial, humanoid life" being created by a mystical adept. The golem is subsequently destroyed without moral concern. In the stories, the golem is not considered to have human status because it lacks the capacity to speak. However, were a human clone to be actually produced, Jewish law would give the clone human status and the Jewish imperative of the preciousness of life would require protection and preservation.

Jewish norms of parenthood and lineage also bear on cloning. The more the processes of parenting—including sexual intercourse, conferral of genetic identity, fetal gestation in a woman's womb, birth, and raising the child—are severed

---

Reprinted with permission from the Program for Ethics, Science, and the Environment, Oregon State University, 1998.

from the actual creation of life, objections from the Jewish tradition will increase. Within an ethic of responsibility, the parent-child relationship provides a basis for reciprocal responsibilities. In the context of cloning (or other reproductive technologies), the sense of responsibility diminishes because it is unclear who has responsibilities to whom. Responsibilities should not be deliberately created and given away.

Judaism also stresses that human diversity is intrinsic to G-d's creation and preservation of the world. A rabbinic maxim supports diversity: "G-d made man from one mold (Adam) yet all the coins (human beings) so minted are different."

Judaism affirms an optimism in the face of scientific uncertainty about unanticipated consequences that is rooted in divine control and care. Indeed, to be too careful and cautious may invite trouble. Thus, human beings do the best that they can. If our best cost/benefit analysis says go ahead, we go ahead. 'G-d protects the simple' is a Talmudic principle that allows us to assume that when we do our best G-d will take care of what we could not foresee or anticipate.

To Clone or Not to Clone?

by Rev. Stanley S. Harakas

Rev. Stanley S. Harakas is Archbishop Iakovos Professor of Orthodox Theology, Emeritus, at Holy Cross Greek Orthodox School of Theology in Brookline, Massachusetts.

ETHICAL STANDARDS DEFINE what "ought to be done" or "what ought not to be done." Such statements will serve to confine the evil, but ethical teachings have never had absolute influence. This allows further questions about what will happen when the unethical is perpetrated. It is these questions that we need to look at now.

Whatever motivations and intentions there might be to take this immoral step, I can think of none that would escape the charge of manufacturing a human being for the purpose of exploiting him or her in a way that depersonalizes the human clone.

Cloning would deliberately deny by design the cloned human being a set of loving and caring parents. The cloned human being would not be the product of love, but of scientific procedures. Rather than being considered persons, the likelihood is that these cloned human beings would be considered "objects" to be used. Given the fallen and sinful condition of our personal and social lives, it is easy to project selfish, greedy, and heartless uses of "manufactured" human clones..

Further, in itself, cloning would violate practically every sacramental di-

---

Reprinted with permission from the Program for Ethics, Science, and the Environment, Oregon State University, 1998.

mension of marriage, family life, physical and spiritual nurture, and the integrity and dignity of the human person. In Orthodox thought, many ethicists are ready to accept technological means to assist a husband and wife to conceive and bear children. We draw the line, however, at the introduction of a third party into that sacred relationship, for it transgresses the spiritual and physical unity of the spouses, blessed by God. How could we approve the substitution of a laboratory for one of the spouses?

Would clones have a soul? One way of clarifying this question is to ask if the clone will have not only intelligence, self-determination, self-consciousness as a person, but will also be able to relate on an inter-personal basis with human beings. There is little doubt that in this sphere, a human clone would have a soul.

If we mean by soul the capacity for relating spiritually to God, given the above, it would seem to me that the clone will be in need of forgiveness, redemption, salvation and sanctification as much as a person born of the mingling of genes which come from two parents.

If genetic material from other animals is added to human DNA, would this make the resulting offspring non-human? How can we answer that question in advance? We do not question the humanity of an existing human person even when an animal organ is transplanted into him or her, such as the experimentations which took place several years ago at Loma Linda University when a baboon heart was transplanted into an infant girl.

It would seem appropriate to correct malformed or deficient DNA with DNA grown in laboratories so long as the purpose is therapeutic, that is, designed to restore human health and normal human functioning.

There is, however, something that provokes a "Star Wars" mentality regarding the creation of a semi-human being. Most responded to the failed Loma Linda experiment with a sense that a boundary had been violated. For human beings to mix human DNA with animal DNA would be, in my judgment, something more than "Playing God." It would be "Playing the Devil."

Who Will Set the Limits?

by Maher Hathout

Maher Hathout is a consultant for the Muslim Public Affairs Council based in Los Angeles.

ISLAM ENCOURAGES RESEARCH and inquiry and places no limits on them. Philosophically, Islam believes that knowledge emanates from Divine sources and human beings have an obligation to interact with this knowledge in order to communicate with God and serve human society. God has taught us what we knew not. He alone has created, creates and will create things we know and things we don't know. He is omnipotent and omnipresent. The Divine teachings explained in the Qur'an exhort Muslims to look at the universe and reflect on the signs created in it.

The Qur'an tells us that God is the best of all creators. Creation can be described in two ways: Creation resulting from putting together things that are already in existence or creating things from nothing. The Qur'an uses the Arabic word *Khaliq* to describe the first type of creation and *Bari* for the second type of creation. The quality of Bari exclusively belongs to God as He alone creates things from nothing.

However, God has empowered human mind to put together things and thus be a *khaliq* of things resulting from this process. For instance, electricity has existed from time immemorial. However, it was the God-given knowledge that human mind employed to use and tame it for the purpose of generating energy. It

---

Reprinted with permission from the Program for Ethics, Science, and the Environment, Oregon State University, 1998.

is a process of creation and human mind is capable of doing it by the will of God.

Thus scientific and empirical investigation is part of human nature as created by God. Any attempt to curb this investigative nature is contrary to Divine principles of creation. Cloning has become a major debate among theologians, scientists and public officials. There are two issues involved in this debate: the research and its application.

A group of theologians and public officials in this country has argued that research on cloning should not be allowed. The Vatican is calling for a ban on cloning humans. Pope John Paul II has denounced dangerous experiments that harm human dignity.

The position of many Muslim scholars is not different than the one adopted by the Vatican. Many of these scholars have missed the point. Research and investigation are part of human nature and they must never be curbed. Human history teaches us that such efforts did not succeed in the past. In more than 50 places in the Qur'an, God invites human beings to reflect, think, research, ponder and work and understand the universe; human beings are then to draw conclusions from their comprehension and adopt a methodology and technology that serves God.

Research pertaining to the cloning of human beings is not an interference in the divine domain of creating things from nothing. It is to understand the dynamics of human life and the process of its creation. It is a manipulation of elements created by God to imitate the creation, not to change it.

However, the moment this research becomes a commodity to be sold and traded like any other commodity, or used for political and cultural superiority, it is a violation of divine principles of serving God and His creation.

Thus, what needs to be curbed is the misuse of the research and technology that emanates from this research. Islam offers strong moral guidelines on the use of research and technology. It prohibits the abuse of any research whether in the natural or social sciences. It lays down clear principles to protect human dignity from any abuse. In other words, Islam makes human mind responsible for human action.

Knowledge Is Not Wisdom

by Viola F. Cordova

Viola F. Cordova, a Native American, teaches philosophy at Lakehead University in Thunder Bay, Ontario, Canada.

THE ABILITY TO enhance the fertility of an infertile female or to develop other means for the re-creation of human beings is based on two assumptions: 1) the needs of individuals, whether real or imagined, are of greater importance than the needs of the larger whole; and, 2) any scientific knowledge adds to the sum of human knowledge and must be pursued regardless of its subsequent social ramifications.

These assumptions are not shared by all human beings. What if there exist assumptions that state that the individual's needs must be determined on the basis of the good of the whole? Or that there is a difference between what can be done and what should be done in the pursuit of knowledge? The first assumption lays a heavier emphasis on the *duties* of the individual to the social whole. The second assumption requires a driving force much different than "knowledge for the sake of knowledge."

The Native American, today, is portrayed as focused primarily on spiritual and mystical endeavors. Absent from this view is the reality that the Native American managed to survive many diverse environments for thousands of years because of a highly pragmatic nature and common views in the midst of linguistic diversity.

Reprinted with permission from the Program for Ethics, Science, and the Environment, Oregon State University, 1998.

The shared views are that individuals exist, first and foremost, as members of a specific cultural group. Secondly, the existence of difference is taken as a natural event. Each group is understood to have a right to exist *as themselves,* as *a group.* Hidden in these views is another assumption: each group exists in a bounded space which may not be expanded. The idea of existence within a bounded space has tremendous implications on the actions of both the group and the individual. Imagine a pebble dropped into a pond: the individual pebble penetrates the water and proceeds to drop to the bottom of the pond; its action, however, creates a far-reaching series of ripples that affects the whole of the pond.

It is the notion of the individual as a pebble dropped into a pond that is the source of the pragmatic nature of the Native American. There are no actions without repercussions. Individual actions have an immediate effect on the group of which the individual is a member and accompanying effects on the surrounding groups. All individual actions are undertaken within the *context* of *his or her* group with the knowledge that the consequences of those actions comprise a "ripple" effect.

In this specific context, the issue of whether manipulation of human genetic material for the purpose of recreating human beings should be pursued takes on a very different meaning. *If* the numbers of human beings were diminishing beyond what it takes for the survival of the group, or the species, there would be no wrong attached to the means used to recreate human beings. The knowledge which leads to such re-creation would be seen as "good" and, therefore, worth pursuing. Given the current circumstances of human population growth and density, the application of the knowledge to clone a human being is unjustified.

We tend to think of "overpopulation" in terms of other nations, China, for example, or India. A quick examination of demographic figures for the United States should quickly change our view: since my childhood in the 1950s there has been a growth in population in the United States of slightly over *one hundred million* persons. Since the growth of population appears to occur exponentially, the outcome of these numbers for the next forty years should astound all of us.

Knowledge, in a Native American sense, is not equated with wisdom. Knowledge with the added awareness of its pragmatic implications comprises wisdom. The ability to clone human beings is certainly a bit of knowledge, but is it wise?

The Opportunity of Cloning

by Ronald Y. Nakasone

Ronald Y. Nakasone, a Buddhist priest, is Professor of Buddhist Studies at the Pacific School of Religion in Berkeley, California.

IAN WILMUT'S SUCCESSFUL cloning of Dolly opens new moral possibilities. The Buddhist response to the possibility of cloning a human being, it seems to me is not if, but when. To be sure many problems will arise and we will need to consider all possibilities and their implications. At the very least we must take into consideration all elements of suffering.

The Buddha anticipated the possibility of unanticipated questions and outlined a fourfold method to respond to new problems. While the problems the Buddha anticipated were rules that regulated the Sangha and its (individual and collective) observance of conduct worthy of the Dharma, his method can be applied to our current interest in human cloning. Essentially, the Buddha said if you are unable to find the answers in the holy texts, use your best judgment.

Vasubandhu, in the ninth chapter of the *Abhidharmakosa sastra* describes the early Indian Buddhist notion of human reproduction. The question of birth is associated with the idea of reincarnation and successive births. Expounding the Sarvastivadan view, Vasubandhu argued in favor of an "intermediate existence" (Skt. *atarabhava,* Jpn. *chuin*), a period that begins with death and continues until one is reborn into another form. An individual who existed in this

---

Reprinted with permission from the Program for Ethics, Science, and the Environment, Oregon State University, 1998.

liminal condition had a correspondingly liminal body. A being in this condition could under the right conditions achieve rebirth. The sight of his future mother would attract this being, if a male; a female would be attracted to the sight of her future father. If the prospective parents happen to be having intercourse on the seventh day after the liminal being's death, the being would then implant itself in the new mother's womb. If the opportunity did not present itself, the liminal being would have another opportunity at the next seven-day cycle. This opportunity is present seven times, until the forty-ninth day before achieving rebirth.

While the notion of an "intermediate state" may have been adequate to explain the transitions between successive lives, the "story" cannot accommodate cloning. The existence of an "intermediate state" has nothing to do with reproducing an identical being. A complete human being would be generated from a single cell; it is the genesis of a new life, not the repackaging of a life in another form. Nor does cloning require a male and a female.

Associated with the idea of successive births is the question of sentience. The traditional definition of sentience is the ability to feel. I am not clear whether sentience means consciousness or awareness of pain or pleasure. Some Buddhists may fudge on this point. The question, thus, becomes: when does consciousness emerge? A ticklish question. However, traditionally Buddhists have avoided the issue by saying that the moment of conception begins a new life.

The current concern of the possibility of human cloning is part, I believe, of a larger consideration of human development and the changes human ingenuity brings to him/herself and the world. Since, for the Buddhist, change is the nature of reality, the questions are how to accommodate change and expand our moral imaginations. Change pushes the boundaries of what we once considered to be the norm. We no longer think it strange or unusual for a child to be conceived through artificial insemination or in vitro fertilization. Medical technology has forced us to expand our moral horizons. We do not think a child conceived in a petri dish to be less than human, although in vitro fertilization bypasses the usual method of human reproduction. The cloning of human beings, like use of artificial insemination and in vitro fertilization, is really about expanding our notion of humanity and our moral parameters.

What does it mean to be human? We are continually pushing the boundaries of what we believe humanity is. New knowledge and the development of new technologies have expanded human possibilities and have added to the meaning of being human. The extension of human longevity from forty-five in 1900 to seventy-eight in 1997, for example, has altered the way in which we think about old age and retirement. In the past only a few individuals lived past fifty. By 2020 more than 20% of the population will be over 65. This demographic shift alters the way society is structured and changes the relationship between generations. Trying to understand this altered social structure and generational relationship forces us to think in new ways.

Would we accord a cloned person the benefits enjoyed by those who are born naturally? I would hope so. It may take time for public sentiment to accept a cloned individual as a person. Is there any law that discriminates against cloned persons? I do not think so. Since in an interdependent world, we rise and fall together as one living body, we have a responsibility to treat everyone with dignity, respect, and gratitude.

When It Comes To Karma...

by Arvind Sharma

Arvind Sharma is Professor of Religious Studies at McGill University in Montreal, Canada.

HINDUISM, ON THE WHOLE, is a religion of options rather than prescriptions, of propositions rather than dogmas; a religion which prefers the article *a* (a truth) to *the* (the truth); a religion of guidelines rather than rules, and a religion which allows for more variations of the basic positions than its own Kama Sutra. As it consists of a frame of mind rather than fixed ideas, its answers are exploratory, rather than catechetical in nature. Within it, *attitude* is more important than *certitude.*

Hinduism will be inclined to resolve the moral issues raised by cloning within its quadripartite axiological framework, built around the values of *Dharma* (Virtue, Morality), *Artha* (Wealth and Power), *Kama* (Aesthetics and Sex) and *Moksa* (Liberation or Salvation). The building blocks are set up as follows:

(1) All these four goals of human endeavour are valid, although the pride of place belongs to *Moksa* in the ultimate analysis;
(2) Wealth and Power and Aesthetics and Sex may be pursued, but subject to virtue or morality;

Reprinted with permission from the Program for Ethics, Science, and the Environment, Oregon State University, 1998.

(3) morality is central to the scheme, as it is the *controlling* value in relation to Wealth and Power, and Aesthetics and Sex; and is the *enabling* value in relation to salvation. In other words, *while permitting cloning as such, Hinduism would insist on its ethical regulation.*

In Indic religions (Hinduism, Buddhism, Jainism), creation is viewed more as a natural cosmic process, a process presided over by God in some forms of the Indic religions and entirely natural in others. Similarly, the range of possible rebirths includes animals and 'angels.' Thus the partition between the natural, the supernatural and the subnatural is thinner than in the Western religions and that open attitude rubs off on the issue of cloning.

Indic religions, although less anthropocentric than the Western ones on the whole, do contain some anthropocentric elements but typically in their *soteriologies* (one stands a better chance of being liberated as a human being). Significantly, however, they are not typically anthropocentric in their cosmologies. Can a clone be liberated or saved?—and if the original is saved is the clone saved as well or vice versa?—are far more interesting questions from their point of view. The answer is 'no' to each question, on the principle that when it comes to Karma or liberation each is on his or her own: clone or no-clone.

Cloning, Ethics, and Religion

by Lee M. Silver

Lee M. Silver is Professor in the Departments of Molecular Biology, Ecology and Evolutionary Biology, and in the Program in Neuroscience at Princeton University.

ON SUNDAY MORNING, 23 February 1997, the world awoke to a technological advance that shook the foundations of biology and philosophy. On that day, we were introduced to Dolly, a 6-month-old lamb that had been cloned directly from a single cell taken from the breast tissue of an adult donor. Perhaps more astonished by this accomplishment than any of their neighbors were the scientists who actually worked in the field of mammalian genetics and embryology. Outside the lab where the cloning had actually taken place, most of us thought it could never happen. Oh, we would say that perhaps at some point in the distant future, cloning might become feasible through the use of sophisticated biotechnologies far beyond those available to us now. But what many of us really believed, deep in our hearts, was that this was one biological feat we could never master. New life—in the special sense of a conscious being—must have its origins in an embryo formed through the merger of gametes from a mother and father. It was impossible, we thought, for a cell from an adult mammal to become reprogrammed, to start all over again, to generate another entire animal or person in the image of the one born earlier.

"Cloning, Ethics, and Religion," by Lee M. Silver, *Cambridge Quarterly of Healthcare Ethics*, vol. 7, no. 2 (1998), 168–72.

This article is based on material extracted from Silver L.M. *Remaking Eden: Cloning and Beyond in a Brave New World*. Avon Books, 1997.

How wrong we were.

Of course, it wasn't the cloning of a sheep that stirred the imaginations of hundreds of millions of people. It was the idea that humans could now be cloned as well, and many people were terrified by the prospect. Ninety percent of Americans polled within the first week after the story broke felt that human cloning should be banned.[1] And while not unanimous, the opinions of many media pundits, ethicists, and policymakers seemed to follow that of the public at large. The idea that humans might be cloned was called "morally despicable," "repugnant," "totally inappropriate," as well as "ethically wrong, socially misguided and biologically mistaken."[2]

Scientists who work directly in the field of animal genetics and embryology were dismayed by all the attention that now bore down on their research. Most unhappy of all were those associated with the biotechnology industry, which has the most to gain in the short-term from animal applications of the cloning technology.[3] Their fears were not unfounded. In the aftermath of Dolly, polls found that two out of three Americans considered the cloning of *animals* to be morally unacceptable, while 56% said they would not eat meat from cloned animals.[4]

It should not be surprising, then, that scientists tried to play down the feasibility of human cloning. First they said that it might not be possible *at all* to transfer the technology to human cells.[5] And even if human cloning is possible in theory, they said, "it would take years of trial and error before it could be applied successfully," so that "cloning in humans is unlikely any time soon."[6] And even if it becomes possible to apply the technology successfully, they said, "there is no clinical reason why you would do this."[7] And even if a person wanted to clone him- or herself or someone else, he or she wouldn't be able to find trained medical professionals who would be willing to do it.

Really? That's not what science, history, or human nature suggest to me. The cloning of Dolly broke the technological barrier. There is no reason to expect that the technology couldn't be transferred to human cells. On the contrary, there is every reason to expect that it *can* be transferred. If nuclear transplantation works in every mammalian species in which it has been seriously tried, then nuclear transplantation *will* work with human cells as well. It requires only equipment and facilities that are already standard, or easy to obtain by biomedical laboratories and freestanding in vitro fertilization clinics across the world. Although the protocol itself demands the services of highly trained and skilled personnel, there are thousands of people with such skills in dozens of countries.

The initial horror elicited by the announcement of Dolly's birth was due in large part to a misunderstanding by the lay public and the media of what biological cloning is and is not. The science critic Jeremy Rifkin exclaimed: "It's a horrendous crime to make a Xerox (copy) of someone,"[8] and the Irvine, California, rabbi Bernard King was seriously frightened when he asked, "Can the cloning create a soul? Can scientists create the soul that would make a being ethical, moral, caring, loving, all the things we attribute humanity to?"[9] The Catholic priest Father Saunders suggested that "cloning would only produce humanoids or androids—soulless replicas of human beings that could be used as slaves."[10] And *New York Times* writer Brent Staples warned us that "synthetic humans would be easy prey for humanity's worst instincts."[11]

Anyone reading this volume already knows that real human clones will simply be later-born identical twins—nothing more and nothing less. Cloned children will be full-fledged human beings, indistinguishable in biological terms from all other members of the species. But even with this understanding, many ethicists, scholars, and scientists are still vehemently opposed to the use of cloning as means of human reproduction under any circumstances whatsoever. Why do they feel this way? Why does this new reproductive technology upset them so?

First, they say, it's a question of "safety." The cloning procedure has not been proven safe and, as a result, its application toward the generation of newborn children could produce deformities and other types of birth defects. Second, they say that even if physical defects can be avoided, there is the psychological well-being of the cloned child to consider. And third, above and beyond each individual child, they are worried about the horrible effect that cloning will have on society as a whole.

What I will argue here is that people who voice any one or more of these concerns are—either consciously or subconsciously—hiding the real reason they oppose cloning. They have latched on to arguments about safety, psychology, and society because they are simply unable to come up with an ethical argument that is not based on the religious notion that by cloning human beings man will be playing God, and it is wrong to play God.

Let us take a look at the safety argument first. Throughout the 20th century, medical scientists have sought to develop new protocols and drugs for treating disease and alleviating human suffering. The safety of all these new medical protocols was initially unknown. But through experimental testing on animals first, and then volunteer human subjects, safety could be ascertained and governmental agencies—such as the Food and Drug Administration in the United States—could make a decision as to whether the new protocol or drug should be approved for use in standard medical practice.

It would be ludicrous to suggest that legislatures should pass laws banning the application of each newly imagined medical protocol before its safety has been determined. Professional ethics committees, institutional review boards, and the individual ethics of each medical practitioner are relied upon to make sure that hundreds of new experimental protocols are tested and used in an appropriate manner each year. And yet the question of unknown safety alone was the single rationale used by the National Bioethics Advisory Board (NBAC) to propose a ban on human cloning in the United States.

Opposition to cloning on the basis of safety alone is almost surely a losing proposition. Although the media have concocted fantasies of dozens of malformed monster lambs paving the way for the birth of Dolly, fantasy is all it was. Of the 277 fused cells created by Wilmut and his colleagues, only 29 developed into embryos. These 29 embryos were placed into 13 ewes, of which 1 became pregnant and gave birth to Dolly.[12] If safety is measured by the percentage of lambs born in good health, then the record, so far, is 100% for nuclear transplantation from an adult cell (albeit with a sample size of 1).

In fact, there is no scientific basis for the belief that cloned children will be any more prone to genetic problems than naturally conceived children. The commonest type of birth defect results from the presence of an abnormal number of chromosomes in the fertilized egg. This birth defect arises during gamete production and, as such, its frequency should be greatly reduced in embryos formed by cloning. The second most common class of birth defects results from the inheritance of two mutant copies of a gene from two parents who are silent carriers. With cloning, any silent mutation in a donor will be silent in the newly formed embryo and child as well. Finally, much less frequently, birth defects can be caused by new mutations; these will occur with the same frequency in embryos derived through conception or cloning. (Although some scientists have suggested that chromosome shortening in the donor cell will cause cloned children to have a shorter lifespan, there is every reason to expect that chromosome repair in the embryo will eliminate this problem.) Surprisingly, what our current scientific understanding suggests is that birth defects in cloned children could occur less frequently than birth defects in naturally conceived ones.

Once safety has been eliminated as an objection to cloning, the next concern voiced is the psychological well-being of the child. Daniel Callahan, the former director of the Hastings Center, argues that "engineering someone's entire genetic makeup would compromise his or her right to a unique identity."[13] But no such 'right' has been granted by nature—identical twins are born every day as natural clones of each other. Dr. Callahan would have to concede this fact, but he might still argue that just because twins occur naturally does not mean we should create them on purpose.

Dr. Callahan might argue that a cloned child is harmed by knowledge of her future condition. He might say that it's unfair to go through childhood knowing what you will look like as an adult, or being forced to consider future medical ailments that might befall you. But even in the absence of cloning, many children have some sense of the future possibilities encoded in the genes they got from their parents. Furthermore, genetic screening already provides people with the ability to learn about hundreds of disease predispositions. And as genetic knowledge and technology become more and more sophisticated, it will become possible for any human being to learn even more about his or her genetic future than a cloned child could learn from his or her progenitor's past.

It might also be argued that a cloned child will be harmed by having to live up to unrealistic expectations placed on her by her parents. But there is no reason to believe that her parents will be any more unreasonable than many other parents who expect their children to accomplish in their lives what they were unable to accomplish in their own. No one would argue that parents with such tendencies should be prohibited from having children.

But let's grant that among the many cloned children brought into this world, some *will* feel badly about the fact that their genetic constitution is not unique. Is this alone a strong enough reason to ban the practice of cloning? Before answering this question, ask yourself another: Is a child having knowledge of an older twin worse off than a child born into poverty? If we ban the former, shouldn't we ban the latter? Why is it that so many politicians seem to care so much about cloning but so little about the welfare of children in general?

Finally, there are those who argue against cloning based on the perception that it will harm society at large in some way. The *New York Times* columnist William Safire expresses the opinion of many others when he says that "cloning's identicality would restrict evolution."[14] This is bad, he argues, because "the continued interplay of genes . . . is central to humankind's progress." But Mr. Safire is wrong on both practical and theoretical grounds. On practical grounds, even if human cloning became efficient, legal, and popular among those in the moneyed classes (which is itself highly unlikely), it would still only account for a fraction of a percent of all the children born onto this earth. Furthermore, each of the children born by cloning to different families would be different from each other, so where does the identicality come from?

On theoretical grounds, Safire is wrong because humankind's progress has nothing to do with unfettered evolution, which is always unpredictable and not necessarily upward bound. H. G. Wells recognized this principle in his 1895 novel *The Time Machine,* which portrays the evolution of humankind into weak and dimwitted but cuddly little creatures. And Kurt Vonnegut follows this same theme in *Galapagos,* where he suggests that our "big brains" will be the cause of our downfall, and future humans with smaller brains and powerful flippers will be the only remnants of a once great species, a million years hence.

As is so often the case with new reproductive technologies, the real reason that people condemn cloning has nothing to do with technical feasibility, child psychology, societal well-being, or the preservation of the human species. The real reason derives from religious beliefs. It is the sense that cloning leaves God out of the process of human creation, and that man is venturing into places he does not belong. Of course, the 'playing God' objection only makes sense in the context of one definition of God, as a supernatural being who plays a role in the birth of each new member of our species. And even if one holds this particular view of God, it does not necessarily follow that cloning is equivalent to playing

God. Some who consider themselves to be religious have argued that if God didn't want man to clone, "he" wouldn't have made it possible.

Should public policy in a pluralist society be based on a narrow religious point of view? Most people would say no, which is why those who hold this point of view are grasping for secular reasons to support their call for an unconditional ban on the cloning of human beings. When the dust clears from the cloning debate, however, the secular reasons will almost certainly have disappeared. And then, only religious objections will remain.

Notes

1. Data extracted from a Time/CNN poll taken over the 26th and 27th of February 1997 and reported in *Time* on 10 March 1997; and an ABC Nightline poll taken over the same period, with results reported in the *Chicago Tribune* on 2 March 1997.
2. Quotes from the bioethicist Arthur Caplan in *Denver Post* 1997; Feb 24; the bioethicist Thomas Murray in New *York Times* 1997; Mar 6; Congressman Vernon Elders in *New York Times* 1997; Mar 6; and evolutionary biologist Francisco Ayala in *Orange County Register* 1997; Feb 25.
3. James A. Geraghty, president of Genzyme Transgenics Corporation (a Massachusetts biotech company), testified before a Senate committee that "everyone in the biotechnology industry shares the unequivocal conviction that there is no place for the cloning of human beings in our society." *Washington Post* 1997; Mar 13.
4. Data obtained from a Yankelovich poll of 1,005 adults reported in *St. Louis Post-Dispatch* 1997; Mar 9 and a Time/CNN poll reported in New *York Times* 1997; Mar 5.
5. Leonard Bell, president and chief executive of Alexion pharmaceuticals, is quoted as saying, "There is a healthy skepticism whether you can accomplish this efficiently in another species." *New York* Times 1997; Mar 3.
6. Interpretation of the judgments of scientists, reported by Specter M, Kolata G. *New York Times* 1997; Mar 3, and by Herbert W, Sheler JL, Watson T. *U.S. News & World Report* 1997; Mar 10.
7. Quote from Ian Wilmut, the scientist who brought forth Dolly, In Friend T. *USA Today* 1997; Feb 24.
8. Quoted in Kluger J. *Time* 1997; Mar 10.
9. Quoted in McGraw C, Kelleher S. *Orange County Register* 1997; Feb 25.
10. Quoted in the on line version of the *Arlington Catholic Herald* (http://www.catholicherald.com/bissues.htm) 1997; May 16.
11. Staples B. [Editorial]. *New York Times* 1997; Feb 28.
12. Wilmut I, Schnieke AE, McWhir J, Kind AJ, Campbell KHS. Viable offspring derived from fetal and adult mammalian cells. *Nature* 1997; 385: 810–13.
13. Callahan D. [op-ed]. *New York Times* 1997; Feb 26.
14. Safire W. [op-ed] *New York Times* 1997; Feb 27.

What's Wrong with Cloning?

by Richard Dawkins

Richard Dawkins is the Charles Simonyi Professor of Public Understanding of Science at Oxford University.

SCIENCE AND LOGIC cannot tell us what is right and what is wrong (Dawkins, 1998). You cannot, as I was once challenged to do by a belligerent radio interviewer, prove logically from scientific evidence that murder is wrong. But you can deploy logical reasoning, and even scientific facts, in demonstrating to dogmatists that their convictions are mutually contradictory. You can prove that their passionate denunciation of X is incompatible with their equally passionate advocacy of Y, because X and Y, though they had not realized it before, are the same thing (Glover, 1984). Science can show us a new way of thinking about an issue, perhaps open our imaginations in unexpected ways, with the consequence that we see our personal Xs and Ys in different ways and our values change. Sometimes we can be shown a way of seeing that makes us feel more favorably disposed to something that had been distasteful or frightening. But we can also be alerted to menacing implications of something that we had previously thought harmless or frivolously amusing. Cloning provides a case study in the power of scientific thinking to change our minds, in both directions.

Public responses to Dolly the sheep varied but, from President Clinton down, there was almost universal agreement that such a thing must never be allowed to happen to humans. Even those arguing for the medical benefits of cloning human tissues in culture were careful to establish their decent credentials, in the most vigorous terms, by denouncing the very thought that adult humans might be cloned to make babies, like Dolly.

But is it so obviously repugnant that we shouldn't even think about it? Mightn't even you, in your heart of hearts, quite like to be cloned? As Darwin said in another context, it is like confessing a murder, but I think I would. The motivation need have nothing to do with vanity, with thinking that the world would be a better place if there was another one of you living on after you are dead. I have no such illusions. My feeling is founded on pure curiosity. I know how I turned out, having been born in the 1940s, schooled in the 1950s, come of age in the 1960s, and so on. I find it a personally riveting thought that I could watch a small copy of myself, fifty years younger and wearing a baseball hat instead of a British Empire pith helmet, nurtured through the early decades of the twenty-first century. Mightn't it feel almost like turning back your personal clock fifty years? And mightn't it be wonderful to advise your junior copy on where you went wrong, and how to do it better? Isn't this, in (sometimes sadly) watered-down form, one of the motives that drives people to breed children in the ordinary way, by sexual reproduction?

If I have succeeded in my aim, you may be feeling warmer towards the idea of human cloning than before. But now think about the following. Who is most likely to get themselves cloned? A nice person like you? Or someone with power and influence like Saddam Hussein? A hero we'd all like to see more of, like David Attenborough? Or someone who can pay, like Rupert Murdoch? Worse, the technology might not be limited to single copies of the cloned individual. The imagination presents the all-too plausible spectre of *multiple* clones, regiments of identical individuals marching by the thousand, in lockstep to a Brave New Millennium. Phalanxes of identical little Hitlers goose-stepping to the same genetic drum—here is a vision so horrifying as to overshadow any lingering curiosity we might have over the final solution to the "nature or nurture" problem (for multiple cloning, to switch to the positive again, would certainly provide an elegant approach to that ancient conundrum). Science can open our eyes in both directions, towards negative as well as positive possibilities. It cannot tell us which way to turn, but it can help us to see what lies along the alternative paths.

Human cloning already happens by accident—not particularly often but often enough that we all know examples. Identical twins are true clones of each other, with the same genes. Hell's foundations don't quiver every time a pair of identical twins is born. Nobody has ever suggested that identical twins are zombies without individuality or personality. Of those who think anybody has a soul, none has ever suggested that identical twins lack one. So, the new discoveries announced from Edinburgh can't be *all* that radical in their moral and ethical implications.

Nevertheless, the possibility that adult humans might be cloned as babies has potential implications that society would do well to ponder before the reality catches up with us. Even if we could find a legal way of limiting the privilege to universally admired paragons, wouldn't a new Einstein, say, suffer terrible psychological problems? Wouldn't he be teased at school, tormented by unreasonable expectations of genius? But he might turn out even better than the paragon. Old Einstein, however outstanding his genes, had an ordinary education and had to waste his time earning a living in the patent office. Young Einstein could be given an education to match his genes and an inside track to make the best use of his talents from the start.

Turning back to the objections, wouldn't the first cloned child feel a bit of a freak? It would have a birth mother who was no relation, an identical brother or sister who might be old enough to be a great grandparent, and genetic parents perhaps long dead. On the other hand, the stigma of uniqueness is not a new problem, and it is not beyond our wit to solve it. Something like it arose for the first in vitro fertilized babies, yet now they are no longer called "test tube babies" and we hardly know who is one and who is not.

Cloning is said to be unnatural. It is of more academic than ethical interest, but there is a sense in which, to an evolutionary biologist, cloning is more natural than the sexual alternative. I speak of the famous paradox of sex, often called the twofold cost of sex, the cost of meiosis, or the cost of producing sons. I'll explain this, but briefly because it is quite well known. The selfish gene theorem, which treats an animal as a machine programmed to maximize the survival of copies of its genes, has become a favored way of expressing modern Darwinism (see, for example, Mark Ridley, 1996; Matt Ridley, 1996). The rationale, in one tautological sentence, is that all animals are descended from an unbroken line of ancestors who succeeded in passing on those very genes. From this point of view, at least when naively interpreted, sex is paradoxical because a mutant female who spontaneously switched to clonal reproduction would immediately be twice as successful as her sexual rivals. She would produce female offspring, each of whom would bear all her genes, not just half of them. Her grandchildren and more remote descendants, too, would be females containing 100 percent of her genes rather than one quarter, one eighth, and so on.

Our hypothetical mutant must be female rather than male, for an interesting reason which fundamentally amounts to economics. We assume that the number of offspring reared is limited by the economic resources poured into them, and that two nurturing parents can therefore rear twice as many as one single parent. The option of going it alone without a sexual partner is not open to males because single males are not geared up to bear the economic costs of rearing a child. This is especially clear in mammals where males lack a uterus and mammary glands. Even at the level of gametes, and over the whole animal kingdom, there is a basic economic imbalance between large, nutritious eggs and small, swimming sperm. A sperm is well equipped to find an egg. It is not economically equipped to grow on its own. Unlike an egg, it does not have the option of dispensing with the other gamete.

The economic imbalance between the sexes can be redressed later in development, through the medium of paternal care. Many bird species are monogamous, with the male playing an approximately equal role in protecting and feeding the young. In such species the twofold cost of sex is at least substantially reduced. The hypothetical cloning female still exports her genes twice as efficiently to each child. But she has half as many children as her sexual rival, who benefits from the equal economic assistance of a male. The actual magnitude of the cost of sex will vary between twofold (where there is no paternal care) to zero (where the economic contribution of the father equals that of the mother, and the productivity in offspring of a couple is twice that of a single mother).

In most mammals paternal care is either nonexistent or too small to make much of a dent in the twofold cost of sex. Accordingly, from a Darwinian point of view, sex remains something of a paradox. It is, in a way, more "unnatural" than cloning. This piece of reasoning has been the starting point for an extensive theoretical literature with the more or less explicitly desperate aim of finding a benefit of sex sufficiently great to outweigh the twofold cost. A succession of books has tried, with no conspicuous success, to solve this riddle (Williams, 1975; Maynard Smith, 1978; Bell, 1982; Michod & Levin, 1988; Ridley, 1993). The consensus has not moved greatly in the twenty years since Williams's 1975 publication, which began:

> This book is written from a conviction that the prevalence of sexual reproduction in higher plants and animals is inconsistent with current evolutionary theory . . . there is a kind of crisis at hand in evolutionary biology. . . .

and ended:

> I am sure that many readers have already concluded that I really do not understand the role of sex in either organic or biotic evolution. At least I can claim, on the basis of the conflicting views in the recent literature, the consolation of abundant company.

Nevertheless, outside the laboratory, asexual reproduction in mammals, as opposed to some lizards, fish, and various groups of invertebrates, has never been observed. It is quite possible that our ancestral lineage has not reproduced asexually for more than a billion years. There are good reasons for doubting that adult mammals will ever spontaneously clone themselves without artificial aid (Maynard Smith, 1988). So far removed from nature are the ingenious techniques of Dr. Wilmut and his colleagues; they can even make clones of *males* (by borrowing an ovum from a female and removing her own DNA from it). In the circumstances, notwithstanding Darwinian reasoning, ethicists might reasonably feel entitled to call human cloning unnatural.

I think we must beware of a reflex and unthinking antipathy, or "yuk reaction" to everything "unnatural." Certainly cloning is unprecedented among mammals, and certainly if it were widely adopted it would interfere with the natural course of the evolutionary process. But we've been interfering with human evolution ever since we set up social and economic machinery to support individuals who could not otherwise afford to reproduce, and most people don't regard that as self-evidently bad, although it is surely unnatural. It is unnatural to read books, or travel faster than we can run, or scuba dive. As the old joke says, "If God had intended us to fly, he'd never have given us the railway." It's unnatural to wear clothes, yet the people most likely to be scandalized at the unnaturalness of human cloning may be the very people most outraged by (natural) nudity. For good or ill, human cloning would have an impact on society, but it is not clear that it would be any more momentous than the introduction of antibiotics, vaccination, or efficient agriculture, or than the abolition of slavery.

If I am asked for a positive argument in favor of human cloning, my immediate response is to question where the onus of proof lies. There are general arguments based on individual liberty against prohibiting anything that people want to do, unless there is good reason why they should not. Sometimes, when it is hard to peer into the future and see the consequences of doing something new, there is an argument from simple prudence in favor of doing nothing, at least until we know more. If such an argument had been deployed against X rays, whose dangers were appreciated later than their benefits, a number of deaths from radiation sickness might have been averted. But we'd also be deprived of one of medicine's most lifesaving diagnostic tools.

Very often there are excellent reasons for opposing the "individual freedom" argument that people should be allowed to do whatever they want. A libertarian argument in favor of allowing people to play amplified music without restriction is easily countered on grounds of the nuisance and displeasure caused to others. Assuming that some people want to be cloned, the onus is on objectors to produce arguments to the effect that cloning would harm somebody, or some sentient being, or society or the planet at large. We have already seen some such arguments, for instance, that the young clone might feel embarrassed or overburdened by expectations. Notice that such arguments on behalf of the young clone must, in order to work, attribute to the young clone the sentiment, "I wish I had never been born because . . ." Such statements can be made, but they are hard to maintain, and the kind of people most likely to object to cloning are the very people least likely to favor the "I wish I didn't exist" style of argument when it is used in the abortion or the euthanasia debates. As for the harm that cloning might do to third parties, or to society at large, no doubt arguments can be mounted. But they must be strong enough to counter the general "freedom of the individual" presumption in favor of cloning. My suspicion is that it will prove hard to make the case that cloning does more harm to third parties than pop festivals, advertising hoardings, or mobile telephones in trains—to name three pet hates of my own. The fact that I hate something is not, in itself, sufficient justification for stopping others who wish to enjoy it. The onus is on the objectors to press a better objection. Personal prejudice, without supporting justification, is not enough.

A convention has grown up that prejudices based upon religion, as opposed to purely personal prejudices, are especially privileged, self-evidently exempt from the need for supporting argument. This is relevant to the present discussion, as I suspect that reflex antipathy to advances in reproductive technology is frequently, at bottom, religiously inspired. Of course people are entitled to their religious, or any other, convictions. But society should beware of assuming that when a conviction is religious this somehow entitles it to a special kind of respect, over and above the respect we should accord to personal prejudice of any other kind. This was brought home to me by media responses to Dolly.

A news story like Dolly's is always followed by a flurry of energetic press activity. Newspaper columnists sound off, solemnly or facetiously, occasionally intelligently. Radio and television producers seize the telephone and round up panels to discuss and debate the moral and legal issues. Some of these panelists are experts on the science, as you would expect and as is right and proper. Others are distinguished scholars of moral or legal philosophy, which is equally appropriate. Both these categories of person have been invited to the studio in their own right, because of their specialized knowledge or their proven ability to think intelligently and express themselves clearly. The arguments that they have with each other are usually illuminating and rewarding.

But there is another category of obligatory guest. There is the inevitable "representative" of the so-and-so "community," and of course we mustn't forget the "voice" from the such-and-such "tradition." Not to mince words, the religious lobby. Lobbies in the plural, I should say, because all the religions (or "cultures" as we are nowadays asked to call them) have their point of view, and they all have to be represented lest their respective "communities" feel slighted. This has the incidental effect of multiplying the sheer number of people in the studio, with consequent consumption, if not waste, of time. It also, I believe, often has the effect of lowering the level of expertise and intelligence in the studio. This is only to be expected, given that these spokesmen are chosen not because of their own qualifications in the field, or because they can think, but simply because they represent a particular section of the community.

Out of good manners I shall not mention names, but during the admirable Dolly's week of fame I took part in broadcast or televised discussions of cloning with several prominent religious leaders, and it was not edifying. One of the most eminent of these spokesmen, recently elevated to the House of Lords, got off to a flying start by refusing to shake hands with the women in the television studio, apparently for fear they might be menstruating or otherwise "unclean." They took the insult more graciously than I would have, and with the "respect" always bestowed on religious prejudice—but no other kind of prejudice. When the panel discussion got going, the woman in the chair, treating this bearded patriarch with great deference, asked him to spell out the harm that cloning might do, and he answered that atomic bombs were harmful. Yes indeed, no possibility of disagreement there. But wasn't the discussion supposed to be about cloning?

Since it was his choice to shift the discussion to atomic bombs, perhaps he knew more about physics than about biology? But no, having delivered himself of the daring falsehood that Einstein split the atom, the sage switched with confidence to geological history. He made the telling point that, since God labored six days and then rested on the seventh, scientists too ought to know when to call a halt. Now, either he really believed that the world was made in six days, in which case his ignorance alone disqualifies him from being taken seriously, or, as the chairwoman charitably suggested, he intended the point purely as an allegory—in which case it was a lousy allegory. Sometimes in life it is a good idea to stop, sometimes it is a good idea to go on. The trick is to decide *when* to stop. The allegory of God resting on the seventh day cannot, in itself tell us whether we have reached the right point to stop in some particular case. As allegory, the six-day creation story is empty. As history, it is false. So why bring it up?

The representative of a rival religion on the same panel was frankly confused. He voiced the common fear that a human clone would lack individuality. It would not be a whole, separate human being but a mere soulless automaton. When I warned him that his words might be offensive to identical twins, he said that identical twins were a quite different case. Why? Because they occur naturally, rather than under artificial conditions. Once again, no disagreement about that. But weren't we talking about "individuality," and whether clones are "whole human beings" or soulless automata? How does the "naturalness" of their birth bear upon that question?

This religious spokesman seemed simply unable to grasp that there were two separate arguments going on: first, whether clones are autonomous individuals (in which case the analogy with identical twins is inescapable and his fear groundless); and second, whether there is something objectionable about artificial interference in the natural processes of reproduction (in which case other arguments should be deployed—and could have been—but weren't). I don't want to sound uncharitable, but I respectfully submit to the producers who put together these panels that merely being a spokesman for a particular "tradition," "culture" or "community" may not be enough. Isn't a certain minimal qualification in the IQ department desirable too?

On a different panel, this time for radio, yet another religious leader was similarly perplexed by identical twins. He too had "theological" grounds for fearing that a clone would not be a separate individual and would therefore lack "dignity." He was swiftly informed of the undisputed scientific fact that identical twins are clones of each other with the same genes, like Dolly except that Dolly's clone is older. Did he really mean to say that identical twins (and we all know some) lack the dignity of separate individuality? His reason for denying the relevance of the twin analogy was even odder than the previous one. Indeed it was transparently self-contradictory. He had great faith, he informed us, in the power of nurture over nature. Nurture is why identical twins are really different individuals. When you get to know a pair of twins, he concluded triumphantly, they even *look* a bit different.

Er, quite so. And if a pair of clones were separated by fifty years, wouldn't their respective nurtures be even *more* different? Haven't you just shot yourself in your theological foot? He just didn't get it—but after all he hadn't been chosen for his ability to follow an argument.

Religious lobbies, spokesmen of "traditions" and "communities," enjoy privileged access not only to the media but also to influential committees of the great and the good, to governments and school boards. Their views are regularly sought, and heard with exaggerated "respect," by parliamentary committees. You can be sure that, if a royal commission were set up to advise on cloning policy, religious lobbies would be prominently represented. Religious spokesmen and spokeswomen enjoy an inside track to influence and power which others have to earn through their own ability or expertise. What is the justification for this? Maybe there is a good reason, and I'm ready to be persuaded by it. But I find it hard to imagine what it could be.

To put it brutally and more generally, why has our society so meekly acquiesced in the idea that religious views have to be respected automatically and without question? If I want you to respect my views on politics, science or art, I have to earn that respect by argument, reason, eloquence, relevant knowledge. I have to withstand counterarguments from you. But if I have a view that is part of my religion, critics must respectfully tiptoe away or brave the indignation of society at large. Why are religious opinions off limits in this way? Why do we have to respect them, simply because they are religious?

It is also not clear how it is decided which of many mutually contradictory religions should be granted this unquestioned respect, this unearned influence. If we decide to invite a Christian spokesman into the television studio or the royal commission, should it be a Catholic or a Protestant, or do we have to have both to make it fair? (In Northern Ireland the difference is, after all, important enough to constitute a recognized motive for murder.) If we have a Jew and a Muslim, must we have both Orthodox and Reformed, both Shiite and Sunni? And then why not Moonies, Scientologists and Druids?

Society accepts that parents have an automatic right to bring their children up with particular religious opinions and can withdraw them from, say, biology classes that teach evolution. Yet we'd all be scandalized if children were withdrawn from art history classes that teach about artists not to their parents' taste. We meekly agree, if a student says, "Because of my religion I can't take my final examination on the day appointed, so no matter what the inconvenience, you'll have to set a special examination for me." It is not obvious why we treat such a demand with any more respect than, say, "Because of my basketball match (or because of my mother's birthday party, etc.) I can't take the examination on a particular day." Such favored treatment for religious opinion reaches its apogee in wartime. A highly intelligent and sincere individual who justifies his personal pacifism by deeply thought-out moral philosophic arguments finds it hard to achieve conscientious objector status. If only he had been born into a religion whose scriptures forbid fighting, he'd have needed no other arguments at all. It is the same unquestioned respect for religious leaders that causes society to beat a path to their door whenever an issue like cloning is in the air. Perhaps, instead, we should listen to those whose words themselves justify our heeding them.

Science, to repeat, cannot tell us what is right or wrong. You cannot find rules for living the good life, or rules for the good governance of society, written in the book of nature. But it doesn't follow from this that any other book, or any other discipline, can serve instead. There is a fallacious tendency to think that, because science cannot answer a particular kind of question, religion can. Where morals and values are concerned, there are no certain answers to be found in books. We have to grow up, decide what kind of society we want to live in and think through the difficult pragmatic problems of achieving it. If we have decided that a democratic, free society is what we want, it seems to follow that people's wishes should be obstructed only with good reason. In the case of human cloning, if some people want to do it, the onus is on those who would ban it to spell out what harm it would do, and to whom.

References

Bell, G. (1982). *The Masterpiece of Nature.* London: Croom Helm.

Dawkins, R. (1998). "The Values of Science and the Science of Values." In J. Ree and C. W. C. Williams (eds.), *The Values of Science: The Oxford Amnesty Lectures* 1997. New York: Westview.

Glover, J. (1984). *What Sort of People Should There Be?* London: Pelican.

Maynard Smith, J. (1978). *The Evolution of Sex.* Cambridge: Cambridge University Press.

Maynard Smith, J. (1988). "Why Sex?" In J. Maynard Smith (ed.), *Did Darwin Get It Right?* London: Penguin.

Michod, R. E., and Levin, B. R. (1988). *The Evolution of Sex.* Sunderland, Mass.: Sinauer.

Ridley, Mark (1996). *Evolution* (second edition). Oxford: Blackwell Scientific Publications.

Ridley, Matt (1993). *The Red Queen.* London: Viking.

Ridley, Matt (1996). *The Origins of Virtue.* London: Viking.

Williams, G. C. (1975). *Sex and Evolution.* Princeton, N.J.: Princeton University Press.

Part III

Perspectives from Philosophy

Perspectives from Philosophy: Introduction

THE PROSPECT OF human cloning elicits a host of philosophical questions. In order to more properly address them, the phenomenon of cloning must be understood within its wider societal context. Have any fundamental differences in the social climate affected how we view technologies? Are we now surrendering to the technological imperative?

Philosophical issues also unfold on various deontological and utilitarian levels. For instance, will cloning lead to objectification and exploitation of the human clone? What are likely consequences of human cloning upon siblings and upon society? The media has played a conspicuous role in fueling the debate and shaping the contours of the discussion. A deeper issue lies in the public's apparent fascination with cloning. Though terrifying for many, why has it boldly seized our attention? Does it relate to our culture's fixation on individualism? In what ways does human cloning challenge our beliefs concerning human individuality and uniqueness?

Human cloning calls into question the nature and extent of moral rights. Can human cloning be appropriately situated within the context of autonomy and individual rights? Do we possess, for instance, a genuine moral right to be genetically unique? And how far should the expression of individual rights go? Does human cloning cause real harm? If so, in what ways would a human clone be harmed? What, in fact, constitutes a real harm?

Philosophical evaluation of the moral legitimacy of human cloning must also consider the reasons for cloning. Are the motives justifiable? For that matter, why do we desire genetically related offspring in the first place? Are our motives for this desire acceptable, or must we assume that the motives behind human cloning are in themselves suspect?

In his brief historical sketch of how the interest in cloning was a major part of the "new biology" in the 1960s, Daniel Callahan points to some subtle though significant differences in the way cloning was viewed then and how it is perceived today. These differences provide a necessary context for the current cloning debate. For instance, the 1960s and 1970s witnessed a substantive intellectual resistance to technology, whereas we now have become more accepting and enthusiastic about the promises of technology, especially those of biotechnologies. Also, reproductive rights are much more prominent today, as Americans now seem to view procreation as a nearly absolute moral right. This attitude affects how we view infertility, and it constitutes a powerful factor in how we evaluate the prospect of human cloning. In the past, infertility was considered a misfortune; today it is increasingly regarded as a pathology. Human cloning's apparent capacity to overcome infertility may therefore become a major factor in its eventual acceptance. In which case, rarer applications, such as for bone marrow donation, may be more easily accepted and eventually become the rule.

According to Callahan, the possibility that human cloning may be more publicly accepted for these reasons is the rationale behind the National Bioethics Advisory Commission's sunset clause. Callahan himself hedges as to whether human cloning is in itself justifiable. Nevertheless, he strongly disfavors both uncritical surrender to the technological imperative and uncritical acceptance of the status quo's notion of "progress."

In abbreviated fashion, Frances Kamm plainly lays out the bare bones of five major quandaries concerning the cloning of human embryos. First, what constitutes the moral status of human embryos? Second, will the technology lead us to clone human embryos purely for the sake of others, in which case clones become objectified? Third, while most critiques center around the clone itself, what about the negative impact this phenomenon may have upon others? Setting standards as to the types of people we desire to have cloned cannot but have some detrimental effect upon those who lack the desirable characteristics.

A fourth concern focuses on the consequences of viewing persons as interchangeable or "substitutable." And Kamm's final apprehension has to do with the possible exploitation of clones for unethical aims, particularly if the cloning technology falls into the wrong hands.

Patrick Hopkins deftly shows how the media have fashioned the cloning issue for the general public, not only framing the contours of the debate, but impressing upon the public what to *think* of as critical problems, particularly the idea that cloning threatens human individuality and uniqueness. This idea strikes deep chords both because of our apprehension about genetic determinism and because of our culture's obsession with individualism to the degree that the need for uniqueness has become a sort of moral right.

Media portrayals of researchers engaged in cloning are skewed in that they showcase scientists driven by egotistic motives. The media also unfairly depict those who intend to clone as being extremely narcissistic, such as the couple who seek to clone a replacement for their lost child. In this vein, media invest cloning with an apocalyptic aura and Brave New World rhetoric. Nevertheless, Hopkins admits that some of the dangers in human cloning are real and raise significant metaphysical questions about human nature, identity, and uniqueness.

Applying Mill's harm principle, Rosamond Rhodes makes the case that human cloning should not be prohibited unless it can be established with certainty that it causes real harm. Her position rests primarily upon claims of individual rights and liberties, as she contends that the rights of relevant parties—the clone, peers of the clone, and the community—would not necessarily be violated. Any alleged right to be genetically unique, she maintains, would in turn lead to the absurd notion that genetic twins' rights have been breached by nature. Finally, Rhodes argues, objections grounded on religious premises may not be imposed upon others without violating those others' rights to religious liberty. Her discussion therefore occurs within the framework of both autonomy and a utilitarian harm principle.

In asserting that overcoming infertility and preventing the transmission of genetic diseases may be legitimate reasons for human cloning, Rhodes offers an insight that is not often voiced. That is, why is it that we do not question the motives of those who desire children through "normal" means but do so automatically when individuals consider more technological methods? Why are not all motives suspect, regardless of method? And indeed, what constitutes a "good" reason for having children?

Moreover, Rhodes cautions, we should not view legislation as the panacea for all controversies. As an alternative, she suggests a waiting period for thoughtful education, discussion, debate, and assessment of outcomes before legislation is enacted. Rhodes therefore advocates a period of tolerant and cautious use of cloning.

As his title indicates, Carson Strong focuses on the benefit that cloning provides for infertile couples. In the process, he tackles some major issues, such as whether or not there are good reasons for desiring genetic offspring. Listing six valid reasons, he isolates two of them as providing equally sound justifications for cloning: the metaphysical dignity in participating in the creation of an individual with self-consciousness and the opportunity for the embodiment of mutual love.

As for the familiar arguments against cloning as a threat to individuality and a source of psychological harm, Strong claims that similar potential harms exist with respect to the more natural ways of reproducing, although the dangers there may be more speculative. He recommends counseling to offset and minimize these potential harms. Another interesting feature in Strong's analysis is his discussion of possible Oedipal and Electral considerations when obtaining a clone of, respectively, the father or the mother. These are legitimate points, as is his comparative evaluation of human cloning and adoption. Against those who point to adoption as the more morally appropriate avenue, Strong cites the difficulties involved in the adoption process as being more burdensome than those encountered in cloning.

At the heart of Strong's argument is his explication of the notion of "harm" and its distinction from "wronging." Citing Feinberg's analysis, he claims that there is no feasible basis for assuming that the clone would be harmed. Echoing many commentators, he states that any wrongs wrought upon the clone would depend upon the intentions and actions of the parents, and he asserts that at this point, there is no empirical evidence that parental motives and actions would wrong the clone's well-being.

Strong contends that human cloning makes moral sense when it comes to relieving infertility, and that for this reason, it ought to be sensibly implemented in policy. He maintains, however, that any such policy should be restricted to infertile couples, and that cloning for other reasons should not be allowed. Recognizing that this poses differing policies for fertile and infertile couples, he concludes that it does not threaten the well-being of fertile couples.

Gregory Pence raises a set of crucial questions regarding the morality of human cloning in particular and asexual reproduction in general. Utilizing, as does Rhodes, John Stuart Mill's famous harm principle, he first questions concerns whether or not real harm is incurred through this exercise of personal liberty. In his second question, Pence again raises the issue of whether harm is incurred by insisting upon a more practical view of morality, asserting that ascertaining both motives and consequences is critical in making a moral assessment of human cloning.

Pence's most penetrating challenge is to the critics of human cloning who unquestioningly assume that the motives for cloning are necessarily bad. Singling out McCormick's position, he points out that such assumptions lack any empirical basis. Finally, Pence attacks slippery-slope arguments against human cloning, and he cites Toffler as a prime example of a disingenuous attempt to hype the repercussions. Applying these four lines of inquiry, Pence argues that the criticisms against human cloning are both weak and indefensible.

Cloning: Then and Now

by Daniel Callahan

Daniel Callahan is the founder and former Director of the Hastings Center in Garrison, New York. He now heads the Center's International Studies Program.

THE POSSIBILITY OF human cloning first surfaced in the 1960s, stimulated by the report that a salamander had been cloned. James D. Watson and Joshua Lederberg, distinguished Nobel laureates, speculated that the cloning of human beings might one day be within reach; it was only a matter of time. Bioethics was still at that point in its infancy—indeed, the term "bioethics" was not even widely used then—and cloning immediately caught the eye of a number of those beginning to write in the field. They included Paul Ramsey, Hans Jonas, and Leon Kass. Cloning became one of the symbolic issues of what was, at that time, called "the new biology," a biology that would be dominated by molecular genetics. Over a period of five years or so in the early 1970s a number of articles and book chapters on the ethical issues appeared, discussing cloning in its own right and cloning as a token of the radical genetic possibilities.

While here and there a supportive voice could be found for the prospect of human cloning, the overwhelming reaction, professional and lay, was negative. Although there was comparatively little public discussion, my guess is that there would have been as great a sense of repugnance then as there has been recently. And if there had been some kind of government commission to study the sub-

"Cloning: Then and Now," by Daniel Callahan, *Cambridge Quarterly of Healthcare Ethics*, vol. 7, no. 2 (1998), 141–44.

ject, it would almost certainly have recommended a ban on any efforts to clone a human being.

Now if my speculation about the situation 20 to 25 years ago is correct, one might easily conclude that nothing much has changed. Is not the present debate simply a rerun of the earlier debate, with nothing very new added? In essence that is true. No arguments have been advanced this time that were not anticipated and discussed in the 1970s. As had happened with other problems in bioethics (and with genetic engineering most notably), the speculative discussions prior to important scientific breakthroughs were remarkably prescient. The actuality of biological progress often adds little to what can be imagined in advance.

Yet if it is true that no substantially new arguments have appeared over the past two decades, there are I believe some subtle differences this time. Three of them are worth some comment. In bioethics, there is by far a more favorable response to scientific and technological developments than was then the case. Permissive, quasi-libertarian attitudes toward reproductive rights that were barely noticeable earlier now have far more substance and support. And imagined or projected research benefits have a stronger prima facie claim now, particularly for the relief of infertility.

1. *The response to scientific and technological developments.* Bioethics came to life in the mid- to late-1960s, at a time not only of great technological advances in medicine but also of great social upheaval in many areas of American cultural life. Almost forgotten now as part of the "sixties" phenomenon was a strong anti-technology strain. A common phrase, "the greening of America," caught well some of that spirit, and there were a number of writers as prepared to indict technology for America's failings as they were to indict sexism, racism, and militarism.

While it would be a mistake to see Ramsey, Jonas, and Kass as characteristic sixties thinkers—they would have been appalled at such a label—their thinking about biological and genetic technology was surely compatible with the general suspicion of technology that was then current. In strongly opposing the idea of human cloning, they were not regarded as Luddites, or radicals, nor were they swimming against the tide. In mainline intellectual circles it was acceptable enough to be wary of technology, even to assault it. It is probably no accident that Hans Jonas, who wrote so compellingly on technology and its potentially deleterious effects, was lauded in Germany well into the 1990s, that same contemporary Germany that has seen the most radical "green" movement and the most open, enduring hostility to genetic technology.

There has been considerable change since the 1960s and 1970s. Biomedical research and technological innovation now encounter little intellectual resistance. Enthusiasm and support are more likely. There is no serious "green" movement in biotechnology here as in Germany. Save possibly for Jeremy

Rifkin, there are no regular, much less celebrated, critics of biotechnology. Technology-bashing has gone out of style. The National Institutes of Health, and *particularly* its Human Genome Project, receive constant budget increases, and that at a time of budget cutting more broadly of government programs. The genome project, moreover, has no notable opponents in bioethics—and it would probably have support *even* if it did not lavish so much money on bioethics.

Cloning, in a word, now has behind it a culture far more supportive of biotechnological innovation than was the case in the 1960s and 1970s. Even if human cloning itself has been, for the moment, rejected, animal cloning will go forward. If some *clear* potential benefits can be envisioned for human cloning, the research will find a background culture likely to be welcoming rather than hostile. And if money can be made off of such a development, its chances will be greatly enhanced.

2. *Reproductive rights.* The right to procreate, as a claimed human right, is primarily of post–World War II vintage. It took hold first in the United States with the acceptance of artificial insemination (AID) and was strengthened by a series of court decisions upholding contraception and abortion. The emergence of in vitro fertilization in 1978, widespread surrogate motherhood in the 1980s, and a continuous stream of other technological developments over the past three decades have provided a wide range of techniques to pursue reproductive choice. It is not clear what, if any, limits remain any longer to an exercise of those rights. Consider the progression of a claimed right: from a right to have or not have children as one chooses, to a right to have them any way one can, and then to a right to have the kind of child one wants.

While some have contended that there is no natural right to knowingly procreate a defective or severely handicapped child, there have been no serious moves to legally or otherwise limit such procreation. The right to procreation has, then, slowly become almost a moral absolute. But that was not the case in the early 1970s, when the reproductive rights movement was just getting off the ground. It was the 1973 *Roe v. Wade* abortion decision that greatly accelerated it.

While the National Bioethics Advisory Commission ultimately rejected a reproductive rights claim for human cloning, it is important to note that it felt the need to give that viewpoint ample exposure. Moreover, when the commission called for a five-year ban followed by a sunset provision—to allow time for more scientific information to develop and for public discussion to go forward—it surely left the door open for another round of reproductive rights advocacy. For that matter, if the proposed five-year ban is eventually to be lifted because of a change in public attitudes, then it is likely that putative reproductive rights will be a principal reason for that happening. Together with the possibility of more effective relief from infertility (to which I will next turn) it is the most powerful viewpoint waiting in the wings to be successfully deployed. If procreation is, as claimed, purely a private matter, and if it is thought wrong to morally judge the means people choose to have children, or their reasons for having them, then it is hard to see how cloning can long be resisted.

3. *Infertility relief and research possibilities.* The potential benefits of scientific research have long been recognized in the United States, going back to the enthusiasm of Thomas Jefferson in the early years of American history. Biomedical research has in recent years had a particularly privileged status, commanding constant increases in government support even in the face of budget restrictions and cuts. Meanwhile, lay groups supportive of research on one undesirable medical condition or another have proliferated. Together they constitute a powerful advocacy force. The fact that the private sector profits enormously from the fruits of research adds still another potent factor supportive of research.

A practical outcome of all these factors working together is that in the face of ethical objections to some biotechnological aspirations there is no more powerful antidote than the claim of potential scientific and clinical benefits. Whether it be the basic biological knowledge that research can bring, or the direct improvements to health, it is a claim difficult to resist. What seems notably different now from two decades ago is the extent of the imaginative projections of research and clinical benefits from cloning. This is most striking in the area of infertility relief. It is estimated that one in seven people desiring to procreate are infertile for one reason or another. Among the important social causes of infertility are late procreation and the effects of sexually transmitted diseases. The relief of infertility has thus emerged as a major growth area in medicine. And, save for the now-traditional claims that some new line of research may lead to a cure for cancer, no claim seems so powerful as the possibility of curing infertility or otherwise dealing with complex procreation issues.

In its report, the bioethics commission envisioned, through three hypothetical cases, some reasons why people would turn to cloning: to help a couple both of whom are carriers of a lethal recessive gene; to procreate a child with the cells of a deceased husband; and to save the life of a child who needs a bone marrow transplant. What is striking about the offering of them, however, is that it now seems to be considered plausible to take seriously rare cases, as if—because they show how human cloning could benefit some few individuals—that creates reasons to accept it. The commission did not give in to such claims, but it treated them with a seriousness that I doubt would have been present in the 1970s.

Hardly anyone, so far as I can recall, came forward earlier with comparable idiosyncratic scenarios and offered them as serious reasons to support human cloning. But it was also the case in those days that the relief of infertility, and complex procreative problems, simply did not command the kind of attention or have the kind of political and advocacy support now present. It is as if infertility, once accepted as a fact of life, even if a sad one, is now thought to be some enormous menace to personal happiness, to be eradicated by every means possible. It is an odd turn in a world not suffering from underpopulation and in a society where a large number of couples deliberately choose not to have children.

What of the Future?

In citing what I take to be three subtle but important shifts in the cultural and medical climate since the 1960s and 1970s, I believe the way has now been opened just enough to increase the likelihood that human cloning will be hard to resist in the future. It is that change also, I suggest, that is responsible for the sunset clause proposed by the bioethics commission. That clause makes no particular sense unless there was on the part of the commission some intuition that both the scientific community and the general public could change their minds in the relatively near future—and that the idea of such a change would not be preposterous, much less unthinkable.

In pointing to the changes in the cultural climate since the 1970s, I do not want to imply any approval. The new romance with technology, the seemingly unlimited aims of the reproductive rights movement, and the obsession with scientific progress generally, and the relief of infertility particularly, are nothing to be proud of. I would like to say it is time to turn back the clock. But since it is the very nature of a progress-driven culture to find such a desire reprehensible, I will suggest instead that we turn the clock forward, skipping the present era, and moving on to one that is more sensible and balanced. It may not be too late to do that.

Moral Problems in Cloning Embryos

by Frances Kamm

Frances M. Kamm is a Professor of Philosophy at New York University.

THE POSSIBILITY OF cloning human embryos raises at least the following five ethical issues (and more, no doubt).

(1) What is the moral status of the human embryo? This question is important because if we freeze cloned embryos for later use, and then decide not to use them, we shall have to decide whether they can be disposed of, and also what their status is in the interim—whether they are private property or individuals with rights.

(2) It has been suggested that we might save cloned embryos in order to use fetal tissue or to have organs for use on the twin who has already developed and develops a medical problem. There are two possible scenarios here. In the first, we use the duplicate embryo to bring into being a person who will continue to live on. Then we have the moral problem of bringing into being a person solely for the sake of using it to improve the life of another person. Whether this is permissible may depend on the sort of life that second embryo to develop would have.

But there is another scenario possible. We use the duplicate embryo to get the tissue or organ we need, but prevent its developing to the point where it is a person. Many think that the point is only reached when capacity for consciousness is present. If we can stop it from becoming capable of consciousness, we shall not be using a person for the sake of another person.

These are comments Professor Kamm made in a news report interview. Reprinted by permission of the American Philosophical Association and the author.

(3) Another forecast prompted by embryo cloning is that we shall allow one embryo to develop, seeing what sort of a person it becomes, and then be able to tell which of the duplicate embryos we want to have developed, either by the same family or by other individuals who would otherwise be infertile. This possibility may be stymied by the fact that nurture as well as nature plays a role in what personality and capacities a person has, and so similar expectations for one embryo and another are unrealistic. But the prospect is nevertheless disturbing. If one standard of ideal personhood develops, this suggests that we prefer not to have people exist who lack its characteristics. Even if multiple standards, expressed by different persons, control decisions about which embryos to develop, this diminishes the appreciation by a family of other types of people. This problem could be obviated if selection of which embryos to develop was on a random basis, perhaps only winnowing out obvious handicaps.

(4) If many possible twins of people in existence could also be brought into existence, there may be a growing sense that people are substitutable for one another. This could lead us to care less about the people that already exist, on the mistaken belief that it doesn't matter so much if they go on, so long as an identical twin can be substituted for them. But people are not substitutable in this way: if you die, it is not much consolation to you that a twin will take your place.

(5) There is a threat that government or corporate control of the process could result in many "duplicate" people used for bad purposes. (For example, many compliant soldiers.) This is a possible result of a "slippery slope" toward disaster. But we should first seek ways to regulate the process so that these disasters do not happen, rather than not use it at all for all the good it can bring.

Bad Copies: How Popular Media Represent Cloning as an Ethical Problem

by Patrick D. Hopkins

Patrick D. Hopkins is Professor of Philosophy at Ripon College in Wisconsin.

The media, perhaps more than any other slice of culture, influence what we think and talk about, what we take to be important, what we worry about. And this was especially true when news of Dolly hit the airwaves and newsstands. Most Americans received training in the ethics of cloning before they knew what cloning was. Media coverage fixed the content and outline of the public moral debate, both revealing and creating the dominant public worries about cloning humans. The primary characterization of cloning as an ethical issue centers around three connected concerns: the loss of human uniqueness and individuality, the pathological motivations of a cloner, and the fear of out-of-control scientists.

WITHOUT HAVING READ a single article, heard a single presentation, or taken a single bioethics class, most Americans have already received training in

---

Patrick D. Hopkins, "Bad Copies: How Popular Media Represent Cloning as an Ethical Problem," *Hastings Center Report*, vol. 28, no. 2 (March–April 1998): 6–13. Reprinted by permission of the Hastings Center and the author.

the ethics of cloning. When the news that scientists had cloned an adult animal hit the airwaves and fiber optic cables of the United States, the public heard for the first time (in a venue other than the movies) that cloning an adult human was possible. But the media stories about cloning were not merely about the procedure. In fact, they were not even predominantly about the procedure. Given more time, teasing, and talk was the story about the morality of cloning. Morality was the real news, and just as the majority of people, including policymakers, got their information on the science and technology of cloning from television and print, they got their information on the ethics of cloning from those same sources. The media instructed us on the major ethical concerns of cloning, its social, religious, and psychological significance, and the motivations behind it. Media coverage fixed the content and outline of the public moral debate, both revealing and creating the dominant public worries about the possibility of cloning humans. It is important then to examine the ethical story the media has told, for being cast much more broadly than academic bioethics debates, it will more widely affect social policy and general attitudes.

Although there are, of course, diverse messages sent through the media, in my investigation of television, magazine, newspaper, and online reports, the primary characterization of cloning as an ethical issue centers around three connected worries: the loss of human uniqueness and individuality, the pathological motivations of anyone who would want to clone, and the fear of "out-of-control" science creating a "brave new world."[1]

Copies and Losing Uniqueness

While many traditional ethical concerns might be generated by cloning—worries about medical risk, the use and loss of embryos, cost and availability, using humans as means—overwhelmingly the media focused on the supposed danger to individuality and uniqueness. This paramount concern about losing our uniqueness (and even our identities) results from anxiety over the status of clones as copies. It is impossible to demonstrate the extent to which the media have fixated on the fear of copies without actually showing the many images and playing the many sound bites, but perhaps at least a sense of this fixation can be conveyed through the following examples:

- A *Time* magazine cover shows an image of the Sistine Chapel, but now there are five identical Adam's hands and the question "Where do we draw the line?" The content page shows an infant's photograph, multiplied by twelve, and the question, "Is this a promising technique or a path to madness?" The spread accompanying the main story shows what appears to be an average middle-class couple with their children, except they have eight identical sons (8 November 1993).
- Another *Time* cover shows two large identical pictures of sheep on a

background of thirty or more smaller copies of the same picture, asking "Will There Ever Be Another You?" The contents page announces the creation of a "carbon-copy." The photo spread introducing the main story shows a coin-operated gumball machine dispensing identical white males by the dozen. A later picture shows identical human bodies dropping out of a test tube (10 March 1997).

- A *Newsweek* cover sports three identical babies in lab beakers. Inside is a picture of Warhol's "The Twenty Marilyns" (10 March 1997).
- *U.S. News & World Report* features a drawing of an ink stamp pressing out copies of babies. An enlargement of the same picture shows one of the baby-copies crying—intimating unhappiness with either being a clone or being cloned (10 March 1997).
- ABC's *Nightline* program opens with this tease: "What if you could make an exact copy of a human being? What if you could make as many as you wanted? You could make a copy of deceased relative. Or a copy of yourself—your perfect organ donor." Then a picture of an angelic baby is multiplied over and over until there are scores of identical infants.

This representation of cloning as a frightening mass production of sameness reflects two powerful and widespread ideas. The first is a belief in genetic determinism. Ordinarily, the common public response on news and talk shows to claims about the genetic determination of violent behavior, or adultery, or even happiness is skepticism or rejection. The reason seems to be reluctance to allow anyone to "get away" with proscribed behavior or to believe that one's own happiness or success is predetermined. It is somewhat odd, then, that the reports on cloning indicate a public belief that a clone will be psychologically identical to his or her donor. As it turns out, however, the media reports contain little evidence that the U.S. public does in fact suddenly believe in genetic determinism. The reports simply assume that it does and then attempt to disabuse the public of its error. But most television and magazine stories engage in a confusing, contradictory bit of double-talk (or double-show). The images and not-very-clever headlines all convey unsettling messages that clones will be exact copies, while inside the stories go to some effort to educate us that clones will not in fact be exact copies.

On the *Nightline* program, which first teased viewers with replicating babies, the reporter asks what it means that scientists could create a genetic copy of him. He says:

> If I expect that baby to become another me, a copy, no way, because he can't live my life, can't have my accidents, my good luck, my bad luck, my experiences. So like all identical twins who start out genetically the same, in spite of the similarities, over time they become very distinct, very different people. Environment counts. It shapes the

> genes, it changes them and creates difference. Says Dr. Francis Collins, head of the government's big project on human genes, "genes can't reproduce an exact copy of a person."

Scenes from the movie *The Boys from Brazil* follow this explanation, and then the summary: "So, no matter what you see in the movies, there's no way my clone could ever be an exact or even a close copy of me. Cloning will never make anybody immortal."

On *The Charlie Rose Show*, Rose discusses the possibility of an infertile couple who want to clone themselves. One guest points out that the child would not be a copy of the parent because that child wouldn't have mom's or dad's experiences. Discussing parents who might want to clone a dying child, another guest argues that much of the ethical debate depends on fundamental misconceptions about what genes actually determine. He says that having a genetic copy might tell you something about the risk of disease, but it will tell you little about what that person will be like as an adult. Thus, these hypothetical parents who want to clone a dying child in order not to lose the child will still in fact lose the child. On the PBS *Newshour*, two interviewees both point out that it is a major mistake to think that a clone would be an exact copy. A later broadcast reiterates that the biggest popular misconception about cloning is that one would get an adult copy of oneself.

Some stories, however, are a bit more confused and ambiguous about their rejection of genetic determinism. In *Time*, Charles Krauthammer writes: "[W]hat Dolly . . . promises is not quite a second chance at life (you don't reproduce yourself; you just reproduce a twin) but another soul's chance at *your* life . . . here is the opportunity to pour all the accumulated learning of your life back into a new you, to raise your exact biological double, to guide your very flesh through a second existence" (10 March 1997, p. 61). But most are very clear in their texts (even while contradicting their stories with images). *Newsweek* says: "[O]n the more profound question of what, exactly, a human clone would be, doubters and believers are unanimous. A human clone might resemble, superficially, the individual from whom it was made. But it would differ dramatically in the traits that define an individual" (p. 55). *U.S. News & World Report* says: "Would a cloned human be identical to the original? Identical genes don't produce identical people . . . Parents could clone a second child who eerily resembled their first in appearance, but all the evidence suggests the two would have very different personalities" (p. 60).

While it is admirable that most reports on cloning try to explain a little basic genetics and try to clarify some of the misconceptions about genetic determinism, it is interesting that most of the comments on determinism are geared toward allaying fears that clones will in fact be exact copies. The push in these remarks is less toward basic genetics education and more toward convincing the public that individual uniqueness is not endangered by cloning.

This concern points to the second prominent idea at work in all those eye-catching pictures and headlines representing cloning as mass photo-copying: that a copy of something is necessarily inferior to the "original" (a term of positive value itself) and that copies often devalue their "originals." Though no one quoted in the cloning reports gave any reason or argument why this would be the case, it is clear from the way copies are characterized that they are metaphysically suspect.

For example, *Time* claims: "Dolly does not merely take after her biological mother. She is a carbon copy, a laboratory counterfeit so exact that she is in essence her mother's identical twin" (10 March 1997, p. 62). The term "counterfeit" here implies that clones as copies are fakes, not as real or legitimate as the original—at least if made by humans. And the anticopy rhetoric gets more passionate. The same issue quotes Jeremy Rifkin saying: "It's a horrendous crime to make a Xerox of someone . . . You're putting a human into a genetic straightjacket" (p. 70). A picture of one of Rifkin's protests in an earlier issue shows people holding signs that say, "I like just one of me" (8 November 1993, p. 69). The existence of human copies is not only interpreted as an assault on individuality, however, but on the very essence of human dignity. A *Time* report on embryo cloning says: "For many, the basic sanctity of life seemed to be under attack." The same issue quotes Germain Grisez, a professor of Christian ethics: "The people doing this ought to contemplate splitting themselves in half and see how they like it" (8 November 1993, p. 69).

On *Nightline*, an interviewee asked about the technology behind cloning says:

> There are certain clear points, though, and one is that we have to use our technology to undergird and to build on human dignity, and human dignity, the dignity of the individual has to be at the center of this discussion and plainly the very idea of cloning introduces a problematic into the notion of human dignity. I mean, this is taking somebody's identity and giving it, at the genetic level, to somebody else. I mean, this is what it's all about . . . Once you start doing it to people, human dignity is in the balance.

U.S. News & World Report informs us that many ethicists believe that the interest in cloning will die away, because: "Making copies, they say, pales next to the wonder of creating a unique human being the old-fashioned way" (10 March 1997, p. 59). This idea implies that clones will lack this highly desired property of uniqueness. These amorphous fears about the existence of genetic copies eating away at human dignity, uniqueness, and individuality even begin to get translated into a right of genetic uniqueness. *Time* quotes Daniel Callahan saying: "I think we have a right to our own individual genetic identity . . . I think this could well violate that right" (8 November 1993, p. 68). In a speech replayed on PBS's *Newshour*, President Bill Clinton raises the worry about uniqueness and copying to an even grander scale: "My own view is that human cloning would have to raise deep concerns given our most cherished concepts of faith and humanity. Each human life is unique, born of a miracle that reaches beyond laboratory science. I believe we must respect this profound gift and resist the temptation to replicate ourselves."

At one and the same time, then, the media showcases, exaggerates, and mitigates concerns that clones will be dignity-damaging, individuality-damaging copies. What none of the reports does, however, is question the assumption that even exact copies would in fact have these deleterious metaphysical, moral, and social consequences for the "original" people who were cloned. Instead, even while defusing *The Boys from Brazil* scenarios, the media shores up a peculiar obsession with uniqueness—pouring the weight of that concept into genetic patterns. The belief promulgated almost seems to be that human value or human dignity is a fixed unity attached to a genetic pattern, a zero-sum game in which copies of the pattern have to divide that value up among themselves. The moral and rhetorical weight attached to this idea is amazing, so much so that even the president characterizes cloning as a sinful "temptation" to replicate ourselves.

One has to wonder if the dominant media message about cloning is not a manifestation of a peculiar American emphasis on individualism. It is assumed that uniqueness is an unquestionable good, a paramount metaphysical virtue (an idea I would expect at least a few twins and triplets to challenge). But no one defends why being unique is better than being one of many. It is easy to imagine, however, the media in another culture with different values never mentioning the worry about copies and the loss of uniqueness. Another culture's magazines might instead focus entirely on medical risks (a topic virtually ignored in U.S. popular coverage). As it is, however, American culture's selective passion for uniqueness is threatened by the realization that humans can be copied biologically. This leads to a vaguely valuative fear that cloning is simply un-American. As *Time* puts it:

> What does the sudden ability to make genetic stencils of ourselves say about the concept of individuality? Do the ants and bees and Maoist Chinese have it right? Is a species simply an uberorganism, a collection of multicellular parts to be die-cast as needed? Or is there something about the individual that is lost when the mystical act of conceiving a person becomes standardized into a mere act of photocopying one? (10 March 1997: 67).

Cloning, *Time* worries, is on the side of robotic insects and communist ideology. Not cloning is on the side of American individualism and Mystery.

As with so many other cases, these ideological alignments lead policymakers to use the law to "protect" us and our conventional understanding of ourselves from the unromantic analyses of science. Announcing a federal moratorium on cloning humans, President Clinton said:

> What the legislation will do is to reaffirm our most cherished belief about the miracle of human life and the God-given individuality each person possesses. It will ensure that we do not fall prey to the temptation to replicate ourselves at the expense of those beliefs . . . Banning human cloning reflects our humanity. It is the right thing to do. Creating a child through this new method calls into question our most fundamental beliefs (*Newshour*).

It is telling that the primary reason for opposing cloning, in both the media and in the words of the chief-of-state, is that copying ourselves challenges our *beliefs* about individuality.

Motivations Behind Cloning

If the dominant ethical issue in cloning coverage was the metaphysical danger posed by copies, it is not surprising that people who desire cloning—who by definition want to copy themselves or others—are considered corrupt or misguided. Of course, there are extraordinarily few people in the world who currently intend to use cloning. After all, the possibility presented itself only recently, and even then it was made clear that human cloning was still a way off. However, in trying to imagine what kind of market cloning might have, the media have repeatedly discussed hypothetical scenarios. One can hardly blame people for trying to think of what uses human cloning might be put to. However, the repeated broadcast and printing of various hypothetical situations has a tremendous influence on how cloning is received—especially when these hypotheticals are laced with moral judgments. Empirically accurate or not, these hypothetical examples travel memetically through the public consciousness, becoming almost paradigmatic.[2] Even before anyone actually requests cloning, we already have a picture of the kind of people who would want it—and it's not flattering. Virtuous motives and human cloning are seen as incompatible. Here are some of the major media examples, in order of their frequency.

The Megalomaniac

This character is drawn from movies, whose clips were shown constantly in the days following the cloning announcement. Scenes from *The Boys from Brazil* flashed onto television screens, showing a plot to clone little Hitlers. Scenes from Woody Allen's *Sleeper*, featuring an attempt to clone an evil leader from his leftover nose, and shots of innocent people fleeing the bloodthirsty T-rex clones of *Jurassic Park* had their time as well. But fiction is frighteningly close to reality, we are told. *Nightline* instructs us that irrespective of the law, some real live fellow with enough money could clone himself if he wanted. *Time* hypothesizes a rich industrialist who has never wanted children but now "with a little help from the cloning lab . . . has the opportunity to have a son who would bear not just his name . . . but every scrap of genetic coding that makes him what he is. Now that appeals to the local industrialist. In fact, if this first boy works out, he might even make a few more" (10 March 1997, p. 70). *Time's* assessment of this situation: "Of all the reasons for using the new technology, pure ego raises the most hackles. It's one thing to want to be remembered after you are gone; it's quite another to manufacture a living monument to ensure that you are. Some observers claim to be shocked that anyone would contemplate such a thing. But that's naïve . . . " (10 March 1997, p. 70). The same issue of *Time* warns of "the ultimate nightmare scenario," which begins: "The Despot will not be coming to the cloning lab today. Before long, he knows, the lab's science will come to him. . . . [he] has ruled his little country for 30 years, but now he's getting old . . . As soon as the technology of the cloning lab goes global—as it inevitably must—his people can be assured of his leadership long after he's gone" (p. 71). *U.S. News & World Report* also blithely informs us, in spite of previously rejecting genetic determinism, that a megalomaniac could decide to achieve immortality by cloning an "heir" (p. 60). Less objectionable but still egomaniacal examples are scattered around—brilliant scientists, great physicians, and famous athletes figure prominently as people who would love to copy themselves, or whom others would love to copy.

The Replacement Child

Usually contrasted to the megalomaniac or egomaniac as a more sympathetic middle-class motivation for cloning is the couple who hopes to "replace" a dying child. Even though *Nightline* host Chris Wallace calls this the "best-case scenario," a guest describes the situation as psychologically dangerous for the child and "horrific." Because it would be hard to say no to such sympathetic parents, we should simply not permit the case to arise. The embryo cloning issue of *Time* asks: "Or what about the couple that sets aside, as a matter of course, a clone of each of their children? If one of them died, the child could be replaced with a genetic equivalent" (8 November 1993, p. 68). *U.S. News & World Report* tells us that one of the most common cloning scenarios ethicists consider is parents cloning a child to replace a dying one (10 March 1997), p. 59). The *New York Times* asks us to consider "the case of a couple whose baby was dying and who wanted, literally, to replace the child" (24 February 1997: B8).

Many of these reports undercut their own efforts at genetic education by implying that the resulting child would in fact be a "replacement" while simultaneously quoting scientists and ethicists arguing against genetic determinism. But the most important aspect of this hypothetical is the idea that cloning is the kind of technology that would appeal to people who are pathologically unable to accept the fact of death. The reluctance to accept their loss leads them to create and use a second child (which they mistakenly see as a replacement) for their own comfort. Using the cloned child in this manner makes parents mild

Kantian villains—creating a child as a means toward their own emotional ends. Interestingly , however, in very few of these discussions is there any mention of parents who already have other children following the death of a child, or even of the most common motivation to have children at all—to make parents' lives fuller and more rewarding. Looked at from a wider angle, it's not clear that these hypothetical parents are much different from any other parents, though they are described as particularly misguided.

The Organ-Donor Cloners

Another step up the ladder of using children as means to an end are those who would want to clone their children or themselves in order to save a life (an existing child's or their own). PBS's *Newshour* informs us that although clinical ethicists agree that it would be wrong to clone humans now, it might be permissible in the future once the safety question has been answered, for example in cases where a family needed a donor for a sick child. *Time* opens its special report with the hypothetical case of parents cloning a child to provide bone marrow for their leukemic daughter, telling us "the parents, who face the very likely prospect of losing the one daughter they have, could find themselves raising two of her—the second created expressly to help keep the first alive" (10 March 1997, p. 67).

In answering their own question of who would want to clone a human in the first place, *U.S. News & World Report says:* "to provide transplants for a dying child" (p. 59). It is not unreasonable, of course, to think that this might be attempted. As we have seen with the Ayalas' bone-marrow case, parents will have other children to save existing ones.[3] But in some reports this admittedly questionable means is rhetorically pushed into vague and scarier scenarios. The *New York Times* quotes Richard McCormick saying: "the obvious motives for cloning a human were 'the very reasons you should not.' " Concerned that people would use cloning to replace dying children or create organ donors, he is also afraid it would tempt people toward eugenic engineering (1 March 1997, p. 10). The very first words in *Newsweek's* story on Dolly are: "[Biologist] Keith Campbell wasn't thinking, really, about rooms full of human clones, silently growing spare parts for the person from whom they had been copied" (p. 53). In short, the supposition that people might clone a biological donor quickly makes its way toward eugenic dystopias, from *Nightline's* "babies produced in batches" to *Time's* intimation of an "embryo factory." In one of the very few cases where a bioethicist actually has space for a significant response to these hypotheticals, Ruth Macklin writes in *U.S. News & World Report*:

> Many of the science-fiction scenarios prompted by the prospect of human cloning turn out, upon reflection, to be absurdly improbable. There's the fear, for instance, that parents might clone a child to have "spare parts" in case the original child needs an organ transplant. But

> parents of identical twins don't view one child as an organ farm for the other. Why should cloned children's parents be any different? . . . Banks stocked with the frozen sperm of geniuses already exist. They haven't created a master race because only a tiny number of women have wanted to impregnate themselves this way. Why think it will be different if human cloning becomes available? (p. 64)

The Last-Chance-Infertile-Couple

Presented as the least objectionable motivation for seeking cloning is the case of the infertile couple who have tried all other treatments. Richard Nicholson, on *Nightline*, says the grotesque scenario of a dictator who wants copies of himself is unlikely. Instead, the more likely scenario is of a young infertile couple who after years of fertility treatment have had a child who is later struck down with meningitis. They know they can't have any more kids so they want to clone a child. *Time* claims that relieving the suffering of infertile couple is the "least controversial" aspect of cloning (8 November 1993, p. 67). *U.S. News & World Report* contrasts the megalomaniac who wants to be cloned to other cases where "adults might be tempted to clone themselves," including "a couple in which the man is infertile [who] might opt to clone one of them rather than introduce an outsider's sperm" (p. 61). While as a response to infertility, cloning may be "less controversial," these reports also strongly suggest that it is the medical status and extreme misfortune of infertile people that might justify the use of an otherwise suspect technology. Cloning is treated only as a last resort for those who have failed in all the obviously better ways of procreating—maintaining cloning as a psychologically and morally inferior method of reproduction.

This summary of motivations for cloning demonstrates the extent to which we are already being trained to suspect anyone who might want to use the technique of pathological, pathetic, or gruesome tendencies. In fact, we have been told implicitly and explicitly that the only motives for cloning adults are vicious. *U.S. News & World Report* tells us "On adult cloning, ethicists are more united . . . In fact the same commission that was divided on the issue of twins was unanimous in its conclusion that cloning an adult's twin is 'bizarre . . . narcissistic and ethically impoverished' " (p. 61).

Brave New Rhetoric

Cloning has not been reported as an unmitigated evil. The potential medical and agricultural benefits are usually mentioned. These benefits, however, are always juxtaposed to the dangers of cloning in alarmist, emotion-packed ways—moderately useful medicines and improvements in animal research versus a "brave new world."

Most people have never read *Brave New World*, but that doesn't matter. The scores of references to *Brave New World* aren't about the book; they are about the trope connected to the book. *Brave New World* is a stand alone reference, image, and warning about dehumanization, totalitarianism, and technology-wrought misery—epitomized and made possible by the technology of cloning. There is no comparable book that praises cloning as a liberating technology. *Brave New World* stands alone, framing the issue as a dichotomy between vaguely helpful medicine and Fordist nightmares of enslaved and manufactured citizens. This easy and morally non-neutral reference was a constant presence in clone reporting—along with more contemporary object lessons.

PBS's *Newshour* jumps from an explanation of cloning to a *Jurassic Park* scene where a cloned T-rex terrorizes humans and then to a picture of a copy of *Brave New World. Nightline* teases their story by saying, "Tonight, cloning, dawn of a brave new world" and later asking if we are "tiptoeing into the brave new world?" *Time* tells us: "A line had been crossed. A taboo broken. A Brave New World of cookie-cutter humans, baked and bred to order seemed . . . just over the horizon. Ethicists called up nightmare visions of baby farming, of clones cannibalized for spare parts" (8 November 1993, p. 65). Another issue warns us that, "The possibilities are as endless as they are ghastly: human hybrids, clone armies, slave hatcheries, 'delta' and 'epsilon' sub-beings out of Aldoux Huxley's *Brave New World* (10 March 1997, p. 61). Yet another *Time* tells us that Neti and Ditto (the embryo-cloned rhesus monkeys) "were not so much a step toward a brave new world as a diversion" (17 March 1997, p. 60). *U.S. News & World Report* warns: "A world of clones and drones, of *The Boys from Brazil* . . . was suddenly within reach" (p. 59). The references continue, including the obligatory Frankenstein comparisons. But only rarely do the assumptions get questioned, as when Bonnie Steinbock remarks on PBS that one misconception about cloning is *The Boys from Brazil* scenario where clones are robotic and easily brainwashed. She says cloning is nothing more than asexual reproduction and people usually act frightened of anything new.

The reference to *Brave New World* in cloning reports is consistent with Valerie Hartouni's analysis of its appearance in other reproductive technology debates. She writes:

> In an otherwise diverse and contesting set of literatures spanning medicine, law, ethics, feminism, and public policy . . . *Brave New World* is a persistent and authoritative presence . . . the work is as frequently invoked only in passing or by title. In either case, the authority and centrality of the text are simply assumed, as is its relevance . . . Whether proffered as illustration, prophecy, or specter, invocations of Huxley's tale clearly function as a kind of shorthand for a host of issues having to do generally with the organization, application, and regulation of these new technologies.[4]

Seeding any discussion of cloning with apocalyptic, slippery slope anxiety, *Brave New World* and its contemporary offspring are treated as warnings by far-sighted social critics more attuned to the dangers of science than naive or misguided scientists. This view of science is part and parcel of brave new rhetoric. Science may hold the answers to many important questions, but it is amoral and dangerous, and the scientists who give their lives to it are treated alternatively as arrogant or naive. Article titles such as *Newsweek's* "Little Lamb, Who Made Thee?" point toward scientists' intrusion on God's power, while at the same time exposing their political simplemindedness by writing:

> The Roslin scientists had no sooner trotted out Dolly than they assured everyone who asked that no one would ever, ever, apply the technology that made Dolly to humans. Pressed to answer whether human cloning was next, scientists prattled on about how immoral, illegal and pointless such a step would be. But as *The Guardian* pointed out, "Pointless, unethical and illegal things happen every day." (p. 57)

Time asks if science has finally "stepped over the line" in embryo cloning and assures us later with Dolly that it indeed has. *Time* then quotes Leon Kass: "Science is close to crossing some horrendous boundaries . . . Here is an opportunity for human beings to decide if we're simply going to stand in the path of the technological steamroller or take control and help guide its direction" (10 March 1997, p. 70). PBS shows President Clinton warning scientists against "trying to play God."

While these hackneyed themes inevitably come up, one aspect of the commentaries appears to be different from other similar discussions. While repeatedly casting science as dangerous, and cloning as something that the "people" should stand up and refuse science permission to do, there is a recurring reluctant admission that science is unstoppable and that human cloning is inevitable. *Newsweek* claims that Dolly's creation offers this lesson: "science, for better or worse, almost always wins; ethical qualms may throw some roadblocks in its path, or affect how widespread a technique becomes, but rarely is moral queasiness a match for the onslaught of science" (p. 59). This uncomfortable acquiescence to science and technology's presumed imperialism occurs again and again. A PBS interviewee says that all efforts to limit and regulate technological progress, including railroads and electricity and gunpowder have failed. Host Jim Lehrer summarizes his point: "So if it's possible to clone human beings, human beings will be cloned." Charlie Rose says that there will always be private money to support this research and that government cannot stop it. The *New York Times* quotes Dr. Lee Silver saying that even if laws were in place to forbid cloning, clinics would crop up: "There's no way to stop it . . . Borders don't matter" (24 February 1997, p. B8). *Time* argues that we will not be able to stop cloning because the medical benefits are immense. *Newsweek* quotes Daniel Callahan saying: "In our society there are two values which will allow anyone to do whatever she wants in human reproduction . . . One is the nearly absolute right to reproduce—or not—as you see fit. The other is that just about anything goes in the pursuit of improved health" (p. 60). The collective message here seems to be that a brave new world is detestable, but may be unavoidable.

The Moral of the Copy

We have been taught a morass of conflicting moral and scientific lessons by the media's public assessment of cloning. But regardless of the consistency of smaller messages, the one idea that surfaces clearly is that we tread on the edge of disaster in attempting to copy ourselves. Though we may at times be comforted by biology lectures telling us that clones are not exact copies, the assumption that exact copies would in fact endanger us in some deep moral sense is very much alive. While no doubt this fear of the copy has a number of sources, I suspect one source is simple the sheer, age-old human desire to think of oneself as metaphysically special, possessing a unique mysterious spark of something that cannot be reduced, measured, or worst of all, copied. But this desire is exactly what science challenges, often unwittingly. If science can figure out enough about a human to be able to copy that human, to create a human, then it really has stepped over the line—but not so much a moral line as a line of privileging self-perception. This is the motivation behind the president's insistence that "each human life is unique, born of a miracle that reaches beyond laboratory science." Of course, this was said in the context of a speech banning federal funds for cloning, but it seems odd that we should make a law forbidding laboratory research if we really believe humans are mystical, mysterious, irreducibly miraculous beings. What could laboratory research do to that kind of being? What would be the point? The point is that cloning itself has its own message, an unsettling message that all good copies teach—the originals are not quite as special or mysterious as they thought.

Notes

1. In particular, I refer to these sources: *Time*, 8 November 1993, 10 March 1997, 17 March 1997; *U.S. News & World Report*, 10 March 1997; *Newsweek*, 10 March 1997; *The New York Times*, 24 February 1997, 25 February 1997, 1 March 1997; PBS's *Newshour* program; PBS's *The Charlie Rose Show*; ABC's *Nightline* program.
2. I already notice in my classes and in other groups that these examples are repeatedly cited as evidence that cloning can be put to no good use. The hypotheticals and the presumed motivations of the characters are treated as certainties.
3. See Ronald Munson, *Reflection and Intervention* (Belmont, Calif.: Wadsworth, 1996).
4. See Valerie Hartounie, "*Brave New World* in the Discourses of Reproductive and Genetic Technologies," in *In the Nature of Things: Language, Politics, and the Environment*, ed. Jane Bennet and William Chaloupka (Minneapolis: University of Minnesota Press, 1993), pp. 86–87.

Clones, Harms, and Rights

by Rosamond Rhodes

Rosamond Rhodes is Director of Bioethics Education and Associate Professor in the Department of Medical Education at Mount Sinai School of Medicine.

AS THE POSSIBILITY of cloning humans emerges on the horizon people are worrying about the morality of using the new technology. They are anxious about the ethical borders that might be crossed when duplicate humans can be produced by separating the cells of a newly fertilized human egg or, in the more distant future, by creating a zygote from an existing person's genetic material. They are apprehensive about eugenics, concerned about creating humans as sources of spare parts for others, uneasy about producing humans without intending to allow them to live and develop, and uncomfortable about using duplicate humans as business ventures.

The religiously inclined are concerned about meddling with the "sanctity of life." As Paul Ramsey has explained, "the value of human life is ultimately grounded in the value God is placing on it . . . [The] essence [of human life] is [its] existence before God and to God, and it is from Him."[1] For believers, cloning sounds dangerously close to playing God, trespassing in His domain, or treading on the sanctity of life.

Those who are sensitive to environmental issues are concerned about meddling with the creation of life. Having witnessed so many problems created by

---

"Clones, Harms, and Rights," by Rosamond Rhodes, *Cambridge Quarterly of Healthcare Ethics*, vol. 4, no. 3 (1995): 285–90. Reprinted with the permission of Cambridge University Press and the author.

the short-sighted use of new technologies, they are concerned with upsetting the delicate balance of nature. They imagine that cloning could have serious implications for limiting reproductive diversity and thus eventually impact on the survivability of the species.

Philosophers have been busy forecasting ethical problems that will be hatched by cloning. For example, in a recent note on the moral problems of cloning, Frances Kamm flagged five concerns, three of which would be unique to cloning:

1. By cloning we could develop a standard (or multiple standards) for an ideal person that could, in turn, diminish the appreciation for other types of people.
2. The availability of genetic multiples could make us careless about those who already exist because they could be replaced like interchangeable (fungible) parts.
3. Government or business could control the cloning process and breed qualities for their purposes, for example, compliant soldiers or workers with great endurance and a high tolerance for monotony.[2]

In response to such concerns coming from so many disparate perspectives there has been a call for a moratorium on the research and a demand for legislation to limit or outlaw the use of cloning techniques. For many the specter of cloning is so awful that they want cloning banned or contained even before it begins. Yet, in the face of these premonitions of disaster this paper will argue that, at least for now, we must resist the movement to proscribe or prohibit cloning. Our society's commitment to liberty requires that we allow individuals to make choices according to their own light, and absent actual substantial evidence that such practices cause serious harm, we are not justified in denying individuals the option.

Liberty

From its inception, our society has embraced the value of liberty. Freedom has been our creed and the foundation for building our government. Although there has been a range of interpretation offered for the concept of liberty, John Stuart Mill's account has been given the greatest weight in moral and political philosophy, and in this discussion, because of the strength of his arguments and the analytic power it yields, I will follow his account. As Mill has explained the commitment, for people who extol liberty, "the sole end for which mankind are warranted, individually or collectively, in interfering with the liberty of action of any of their number is self-protection."[3] This principle of limiting legislation, which has become known as the "harm principle," demands that no action be forbidden unless it can be shown to cause harm to others in the enjoyment of their rights.[4]

> As soon as any part of a person's conduct affects prejudicially the interests of others, society has jurisdiction over it, and the question whether the general welfare will or will not be promoted by interfering with it becomes open to discussion. But there is no room for entertaining any such question when a person's conduct affects the interests of no person besides himself, or need not affect them unless they like.

Although anything one person does may give another affront, upset, or sadness and thereby cause some harm, only those actions that "violate a distinct and assignable obligation to any person or persons"[5] may be proscribed by legislation.

Clones, Rights, and Harms

With respect to reproduction by cloning, the question relevant to Mill's criterion is whether anyone's rights would be violated by cloning humans. To answer we must consider all of those anticipated to be violated. As far as I can foresee, those who might be harmed by the production of cloned offspring would include the perspective cloned children, their peers (according to Kamm's scenario), and those in the community who would be upset by people overstepping the line into God's domain. None of these, however, would suffer any violation of rights.

Obviously multiple children with the same genetic inheritance produced by cloning would not be genetically unique. Having a genetic twin, or several of them, however, is neither a clear harm nor a clear benefit. It might be psychologically harmful because genetic twins could be so easily confused or compared. But it could also be psychologically beneficial because of the special sharing, support, and intimacy that might develop between the genetically identical individuals. Further, although cloned duplicates are genetically the same, identical twins or other children from a multiple birth and, theoretically even some natural siblings, share common genetic material. No one has ever charged that their parents had violated their rights by having more than one child with the same DNA. Apparently, even though most humans happen to be genetically unique, no one has a right to be unique. Therefore, no one who produces multiple children by cloning should be prohibited from doing so because of violating the rights of others.

Kamm's worry that we could develop a standard for an ideal person that would diminish our appreciation for other types of people does not meet Mill's standard for prohibiting the choice either. First, it seems that no one has a right to prevent the existence of others who might be superior to themselves and, just by living, make the inferior feel unappreciated. That cloning technology might be the means to enable some superior individuals to be born does not, therefore, violate anyone else's rights. Second, Kamm's concern with the development of a new "ideal" depends on a large and significant number of people being produced by cloning. Because the cost, inconvenience, discomfort, and loss of privacy entailed by the procedure would be likely to make cloning a rarely employed technology, and because there would be a variety of motivations and procreators, the number of individuals produced by cloning would not be great enough or similar enough to have any significant impact on the social ideal of a person. Likewise, other objections to producing large numbers of cloned individuals could be discounted for now because that prospect is not presently being considered.

The religious concern over cloning interfering with the sanctity of life also fails to meet Mill's criterion for legislating against a practice. Although liberty allows individuals the freedom to choose a religious perspective and the freedom to live according to the religious views they embrace, it limits individuals' infringement on the similar rights of others. In other words, no one may impose his own religious views on others. So although no one has the right to interfere with anyone else's religious practice, the others he respects have no right to intervene in his living by his own religious or nonreligious standards. The religious liberty guaranteed by the harm principle does not extend rights to control the lives of others, and so those whose religious sensitivities are upset by the prospect of other people meddling with the creation of human life cannot claim that harm as grounds for limiting others' procreative practice.

The Cautious Skeptic's Objection

Cautious skeptics, who might otherwise be guided by Mill's harm principle, nevertheless object to cloning because they can see no good reason for resorting to cloning technology.[6] Being hesitant about plunging into uncharted territory in the face of well-known omens of disaster, they plead for slowing down the momentum of technological capability. Skeptics imagine egoists using cloning technology to reproduce themselves and, limited only by the extent of their wealth and narcissism, cloning themselves many times. This consideration, which invokes neither harms nor rights, nevertheless deserves a response.

First, there are "good" reasons for employing cloning technology. It could be utilized to manage both fertility and genetic problems. For instance, women with few remaining eggs who want to preserve their chance of having one or more biologically related offspring could want to employ the technology. Because in the most competent programs there is only a one in four chance of producing a pregnancy by *in vitro* fertilization, cloning would allow these women to increase their likelihood of having children. For couples who want to avoid passing on serious genetic abnormalities, cloning genetically screened embryos could allow them the opportunity to increase their likelihood of reproducing without passing along inherited diseases.

Putting these good reasons aside, it is important to point out that we do not question the reasoning that motivates nontechnology-assisted reproduction. The ordinary desire to have biologically related offspring is not challenged even in the face of overpopulation and the large numbers of orphaned children around the world. Without aid and without society's interference people have children to pass along their genes, pressure a partner into marriage, get an apartment, keep a marriage together, get an inheritance, have a real live doll to play with, or have somebody to love. It is not even clear which reasons are good reasons and which are not. But it is clear that privacy and respect for autonomy require that people be allowed to follow their own reasons. And, at least since Hobbes' writings in the seventeenth century, it has been understood that law could only govern action and not thought or belief. Considering the skeptics' objections to cloning therefore leads to limiting social impediments to cloning rather than supporting its prohibition.

An Alternative Approach

In America, legislation has recently become the most popularly advocated solution to every problem of managing medical technology. This may be attributable to the new social awareness of medical needs that has accompanied Clinton's putting access to healthcare on the political agenda. But whatever the reasons, whenever some medical misconduct is brought to light, the announcement is immediately accompanied by calls for legislation to correct the problem. For example, the recent revelation of the radiation experiments of the 1950s was followed by calls for laws to more carefully govern such research. The demand for more laws came in spite of the fact that the presently existing institutional review board (IRB) requirements, which went into effect after these unacceptable clinical trials had occurred, would have been adequate to have prevented those outrageous abuses of human subjects. Then again, according to the emerging pattern, just last March when a UCLA study of treatment for schizophrenia was publicly criticized, the discussions again called for more legislation to protect research subjects even though most of the critical commentary pointed to problems of inadequate implementation of existing rules.[7] It seems that we have become like the fellow with a hammer who sees a nail as the solution to every problem. We have developed a practice of leaping at every situation with legislation before the need for legislation has been demonstrated and before we have adequate empirical evidence to form the basis of desirable and effective laws. Legislation need not be proactive; there are other alternatives.

Two examples illustrate a better approach to the development of social policy related to medicine. One comes from The Netherlands, the other from the history of American nursing. In The Netherlands today, although euthanasia is still officially prohibited under old laws against murder, the society is carrying on a program for deciding whether or not to accept the practice as they also try out standards for regulating it. They began in the late 1980s by formulating guidelines to govern physician-assisted dying. Then they undertook studies to gather empirical evidence on what was in fact being done to whom, when, and how, and to see how that practice was changing over time. One was an official government study, the Remmelink Commission Report of 1991–1992, the other an independent research project by Gerrit van der Wall, 1991–1992. They are now in the process of assessing their study findings so that the guiding policy can be amended based on what they have learned. Eventually they expect to draft a law that will regulate euthanasia but only after they have tried to assess the actual harms that they want to avoid.[8] As Mill advised, their ultimate decisions will be informed by "different experiments of living."[9]

In 1898, Isabel Hampton Robb, the first president of the American Nursing Association advised the association to go slowly in pursuing one of its stated goals, "to establish a code of ethics." She argued that

> it will be better to wait to learn the mind of the greater number on what shall constitute our national code of ethics. This code should be formulated to meet our own special needs in our own special way.

First the American Nurses Association Committee on Ethical Standards worked on developing a professional statement of ideals. Then it took time to gather content from nurses, reevaluate its ideals, and formulate the wording. The Committee finally recommended a code for nurses at the annual convention in 1926. The code was then sent back to the Committee for further work and the suggested code was published in the *American Journal of Nursing* so that further comments could be elicited. The code was ultimately accepted in 1956. (Subsequently it was revised in 1975, 1985, and again in 1989.)[10]

Conclusion

This 58-year history of the development of a nursing code of ethics illustrates again the alternative model of preceding legislation with a period of thoughtful deliberation, gathering empirical data, and assessing the real need for rules. Were such a model to be followed in guiding cloning practice, the process would begin by allowing physicians and their patients to use the new technology directed by the existing recommendations for the employment of technologies (like assessing genetic risk[11]) in "different experiments of living." Existing guidelines typically include requirements to establish precautions against conflict of interest and to provide for the informed consent of participants that would require giving the person information about the risks, benefits, efficacy, and alternatives. Surrogacy programs also typically include psychological screening for participants. Obviously, such safeguards should be part of any program using cloning technology. Counseling and screening practices, however, should be seen as educational enhancements of liberty rather than intrusions on it. Again citing Mill, "human beings owe to each other help to distinguish the better from the worse, and encouragement to choose the former and avoid the latter."[12] Prospective patients being counseled about cloning should be encouraged to seriously evaluate all of the caveats raised by the objectors and to examine their own motives for pursuing the technology. Psychological implications, not only for the resulting children but also for the parents who may be disappointed by their offsprings' lack of success or jealous of their superior achievements, should be raised. But the primary goal for the early stages of employing the new technology should be gathering the empirical evidence about the harms that might ensue and trying to learn how best they might be avoided.

The arguments above all suggest a single conclusion. Instituting legislation to bar or limit the employment of cloning technology would seriously violate our commitment to liberty. Similarly, panels comprised of representatives of particular political agendas that issue constraining policies that actually function as law that circumvents the democratic process, and also the tyranny of the majority and the blaring voice of vocal minorities, can all be enemies of liberty. Instead of arresting advances in cloning technology, those who are uneasy about proceeding with cloning must be tolerant of cloning. We can begin to experiment with rules that would help avoid any harmful consequences. And in the meanwhile, those engaged in research and early applications of the new technology must also study the consequences and guidelines of research and practice in related areas of reproductive medicine. Conscientious care is required in scientific research and for ethical implementation of new technology. And just as scientific advance must be based on careful evaluation of the relevant data, ethical rules that reflect our strongest moral intuitions about the value of liberty must also be based on the experiential data of actual harms and violations of rights.

Notes

1. Ramsey P. The morality of abortion. *Life or Death: Ethics and Options.* Seattle: University of Washington Press, 1968: 72–3.
2. Kamm F. Moral problems in cloning embryos. *American Philosophical Association Newsletter on Philosophy and Medicine* 1994; Spring: 91.
3. Mill JS. *On Liberty.* Indianapolis, Indiana: Hackett Publishing Company, 1978: 9.
4. See note 3. Mill. 1978: 73–4.
5. See note 3. Mill. 1978: 74.
6. The skeptics' argument was forcefully presented to the author by Joe Fitschen in conversation.
7. *New York Times,* 1994; March 10: A1, B10.
8. This discussion draws on a paper by MP Battin, Seven (more) caveats concerning the discussion of euthanasia in The Netherlands. *American Philosophical Association Newsletter on Philosophy and Medicine* 1993;92:76–80.
9. See note 3. Mill. 1978:54.
10. This section follows a paper by E Murphy. A history of nursing ethics, delivered at the Oxford-Mount Sinai Conference, New College, Oxford, April 6, 1994.
11. Committee on Assessing Genetic Risks, Division of Health Sciences Policy, Institute of Medicine, Assessing genetic risks: implications for health and social policy. Washington, DC, 1993.
12. See note 3. Mill. 1978: 74.

Cloning and Infertility

by Carson Strong

Carson Strong is Professor of Philosophy in the Department of Human Values and Ethics at the University of Tennessee in Memphis.

ALTHOUGH THERE ARE important moral arguments against cloning[1] human beings, it has been suggested that there might be exceptional cases in which cloning humans would be ethically permissible.[2–3] One type of supposed exceptional case involves infertile couples who want to have children by cloning. This paper explores whether cloning would be ethically permissible in infertility cases and the separate question of whether we should have a policy allowing cloning in such cases. One caveat should be stated at the beginning, however. After the cloning of a sheep in Scotland, scientists pointed out that using the same technique to clone humans would, at present, involve substantial risks of producing children with birth defects.[4–6] This concern over safety gives compelling support to the view that it would be wrong to attempt human cloning now. Thus, we do not reach the debate about exceptional cases unless the issue of safety can be set aside. I ask the reader to consider the possibility that in the future, humans could be cloned without a significantly elevated risk of birth defects from the cloning process itself. The remainder of this paper assumes, for sake of argument, that cloning technology has advanced to that point. Given this assump-

"Cloning and Infertility," by Carson Strong, *Cambridge Quarterly of Healthcare Ethics*, vol. 7, no. 2 (1998), 279–93.

tion, would cloning in the infertility cases be ethically permissible, and should it be legally permitted?

An example of the type of case in question is a scenario in which the woman cannot produce ova and the man cannot produce sperm capable of fertilizing ova.[7] Like many couples, they wish to have a child genetically related to at least one of them. One approach would use sperm and ova donated by family members, but suppose that no family donors are available in this case. Let us assume, in other words, that cloning is the only way they could have a child genetically related to one of them. Imagine the couple asking their infertility doctor to help them have a child by cloning. This would involve replacing the nucleus of a donated ovum with that of a cell taken from either member of the couple. Suppose that an ovum donor is available who is willing to participate in this process. The infertile couple would decide whether to duplicate genetically the woman or the man. They could try to have a girl or a boy or possibly a child of each sex—fraternal twins. They could use cloning again to have subsequent children: perhaps one the opposite sex of a first child; or another the same sex; or twins again, among other possibilities.

Many would consider cloning ethically unjustifiable in such cases. Following the birth of Dolly, the sheep clone, the response from ethicists, politicians, and journalists was overwhelmingly against cloning human beings.[8–12] In fact, few issues in bioethics seem to have reached the high level of consensus found in our society's opposition to human cloning. This opposition rests on a number of concerns, religious and secular. I will focus on the secular arguments, which include, at least, the following main ones. First, the persons produced would lack genetic uniqueness, and this might be psychologically harmful to them. Second, this reproductive method transforms babymaking into a process similar to manufacturing. Children would become products made according to specification; this would objectify children and adversely affect parental attitudes toward children and other aspects of parent-child relationships. Third, additional abuses might occur if this technology were obtained by totalitarian regimes or other unscrupulous persons.

My main thesis is that the ethics of cloning is not as clear-cut as many seem to think. Specifically, when the arguments against cloning are applied to infertility cases like the one described above, they are not as strong as they might initially appear. Such cases can reasonably move us away from the view that cloning humans is always wrong. Moreover, the arguments for legally prohibiting cloning in such cases are not strong enough to support such restrictions.

Whether cloning in the above case is ethically justifiable rests on the following question: Which is weightier, infertile couples' reproductive freedom to use cloning or the arguments against cloning humans? To address this question, we need to examine closely both the importance of the freedom of infertile couples to utilize cloning and the arguments against its use.

Freedom of Infertile Couples to Use Cloning

Let us begin by asking why the freedom of infertile couples to use cloning should be valued. Because a main reason to use cloning in the above case is to have children who are genetically related to at least one member of the couple, we need to ask whether reasons can be given to value the having of genetically related children. It is worth noting that studies[13–18] have identified a number of reasons people actually give for having genetically related offspring, some of which seem selfish and confused. For example, some people desire genetic children as a way to demonstrate their virility or femininity. The views on which these reasons seem to be based—that virility is central to the worth of a man, and that women must have babies to prove their femininity—are unwarranted. They stereotype sex-roles and overlook ways self-esteem can be enhanced other than by having genetic offspring. Another example involves desiring a genetic child in order to "save" a shaky marriage. This reason fails to address the sources of the marital problems, and the added stress of raising the child might make the marital relationship even more difficult. Some commentators seem to think that the desire for genetic offspring is always unreasonable, as in these examples.[19] Rather than make this assumption, we should consider whether there are defensible reasons that could be given for desiring genetic offspring.

I have explored this question elsewhere, focusing on a category of procreation commonly referred to as "having a child of one's own," sometimes stated simply as "having a child" or "having children."[20] Although these expressions can be interpreted in several ways, I use them to refer to begetting, by sexual intercourse, a child whom one rears or helps rear. This, of course, is the common type of procreation, in which parents raise children genetically their own. My strategy was to try to understand why having genetic offspring might be meaningful to people in this ordinary scenario, and then use this understanding to address assisted reproductive technologies. For the common type of procreation, I identified six reasons people might give for valuing the having of genetic offspring. Briefly, they are as follows: having a child involves participation in the creation of a person; it can be an affirmation of a couple's mutual love and acceptance of each other; it can contribute to sexual intimacy; it provides a link to future persons; it involves experiences of pregnancy and childbirth; and it leads to experiences associated with child rearing. However, the above infertility case differs in several ways from the ordinary type of procreation, including the fact that children would not be created by sexual intercourse. Because of these differences, not all the reasons that might be given in justifying the desire for genetic offspring in the common scenario would be strong reasons in the cloning situation. Nevertheless, I believe that at least the following two reasons would be significant.

Participation in the Creation of a Person

When one 'has a child of one's own,' as defined above, a normal outcome is the creation of an individual with self-consciousness. Philosophers have regarded the phenomenon of self-consciousness with wonder, noting that it raises perplexing questions: What is the relationship between body and mind? How can the physical matter of the brain give rise to consciousness? It is ironic that although we have difficulty giving satisfactory answers to these questions, we can create self-consciousness with relative ease. Each of us who begets or gestates a child who becomes self-conscious participates in the creation of a person. One might say that in having children we participate in the mystery of the creation of self-consciousness. For this reason, among others, some might regard creating a person as an important event, perhaps one with metaphysical or spiritual dimensions (p. 114).[21] Perhaps not all who have children think about procreation in these terms, but this is a reason that can be given to help justify the desire for genetic offspring.

Similarly, the infertile couple might reasonably value the use of cloning because it would enable them to participate in the creation of a person. The member of the couple whose chromosomes are used would participate by providing the genetic material for the new person. Regardless of whose chromosomes are used, if the woman is capable of gestating, she could participate by gestating and giving birth to the child.

It might be objected that the infertile couple could participate in the creation of a person by using donor gametes or preembryos, in the sense that they would authorize the steps taken in an attempt to create a person. Also, if the woman were the gestational mother and used donor gametes or preembryos, then she would participate biologically in the creation of the person. In reply, although these would constitute types of participation, a more direct involvement would occur if one member of the couple contributed genetically to the creation of the child. From the body of one of them would come the makeup of the new person. Cloning would be the only way that the man, in fact, could participate biologically in the creation of the person. This more direct involvement would increase the degree to which the couple participates in the creation of a person, and for some this greater participation might be especially meaningful.

Affirmation of Mutual Love

In the ordinary type of procreation, intentionally having children can be an affirmation of a couple's mutual love and acceptance of each other. It can be a deep expression of acceptance to say to another, in effect, "I want a child to come forth from your body and mine." In such a context there might be an anticipation that the bond between the couple will grow stronger because of common children to whom each has a biological relationship. To intentionally seek the strengthening of their personal bond in this manner can be a further affirmation of mutual love and acceptance.

In the infertility case in question, if cloning is used, then the child would not receive genes from both parents. Nevertheless, a similar affirmation of mutual love is possible if the woman is capable of being the gestational mother and the man's genes are used. In that situation, it remains true that the child comes forth from their two bodies. Assuming mutual love, the woman bears a child having the genes of the man who loves her and is loved by her. Alternatively, suppose that the woman's genes are used. The man then can become the social father of a child having the genes of the woman who loves him and is loved by him; to seek to become social parents in this manner can also be an affirmation of mutual acceptance.

It might be objected that having children by donor gametes or preembryos—or even adopting children—can also be an affirmation of mutual acceptance, for each member of the couple selects the other to be a social parent of their children. These types of affirmation could enrich the couple's relationship with each other. In response, although these would indeed be forms of affirmation, they can be viewed as different from the affirmation involved in trying to have genetically related children. Intentionally to create a child having the partner's genes might be regarded by some as a special type of affirmation, one that would enrich the couple's relationship in its own distinctive way. For some couples, this type of affirmation might have special significance.

In stating these two reasons, I do not mean to imply that one *ought* to desire genetic offspring, much less that one ought to desire cloning if necessary to have genetically related children. Rather, the point is that the desire for genetic offspring—and hence the desire for cloning in the situation being considered—could be supported by reasons that deserve consideration. Although not everyone in the infertile couple's situation would want to pursue cloning, some might. These reasons also help explain why *freedom* to use cloning to have biological children might be considered valuable; namely, because some couples might value either the opportunity to participate directly in the creation of a person or an affirmation of mutual love that can be associated with that endeavor, or both.

Arguments against Cloning

Let us turn to the considerations against cloning, beginning with the arguments based on lack of genetic uniqueness. What exactly are the adverse effects envisioned for persons who are genetically identical to others? Perhaps the most obvious concerns involve the possibility of being one of *many* genetically identical persons—perhaps one of hundreds of clones, or thousands, or even more. One argument is that the clones would be psychologically harmed, in that they would feel insignificant and have low self-esteem. If I know that I am one of many who physically are exact duplicates, then I might easily believe that there is little or nothing special about me. Apart from what the clones would feel, objections to multiple cloning can also be made from a deontological perspective; it would be an affront to their human dignity to be one of so many genetic replicas. There would also be a serious violation of personal autonomy if the clones were under the control of those who produced them.

Lack of genetic uniqueness can raise concerns even if there are not many clones. Imagine, for example, being the single clone of a person 30 years older. There might be a tendency for the older person's life to be regarded as a standard to be met or exceeded by the younger one. If the clone feels pressured to accept that standard, this might be a significant impediment to freedom in directing one's own life. In addition, knowing that one is a clone might be psychologically harmful in this situation too. For example, self-esteem could be diminished; perhaps the child would regard herself as nothing more than a copy of someone who has already traveled the path ahead.

Although these arguments initially appear persuasive, we need to consider the conclusions reached when we apply them to the infertility case. To begin, the arguments based on large cohorts of clones would not be relevant. The couple might create only one or two children using cloning; thus, their use of cloning can be distinguished from scenarios in which large numbers of clones are produced.

The argument that the parent's life might be regarded as a standard would be relevant, but a response can be given. For one thing, parents' lives often are held up as standards, even in the absence of cloning. This can be either good or bad for the child, depending on how it is handled; it has the potential to promote as well as inhibit development of the child's talents, abilities, and autonomy. Similarly, a clone's being given a role model or standard is not necessarily bad. It depends on how the standard is used and regarded by those directly involved. If it is used by parents in a loving and nurturing manner, it can help children develop their autonomy, rather than inhibit it.

It might be objected that when the child is genetically identical to the parent, there will be a tendency for the parent to be less forgiving when the child fails to meet expectations. If so, the parental standard might tend to thwart rather than promote the child's developing autonomy. In reply, it should be noted that this concern is rather speculative. We do not know the extent, if any, to which the child's being genetically identical would tend to promote a domineering attitude on the part of the parent. Moreover, there is a way to address this concern other than forbidding the infertile couple to use cloning; in particular, the couple could be counseled about the possible psychological dimensions of parenthood through cloning. Psychological counseling already is widely accepted in preparing infertile couples for various noncoital reproductive methods, such as donor insemination and surrogate motherhood. Similarly, it would be appropriate to offer psychological counseling if cloning is made available to infertile couples. The aims of such counseling could include raising awareness about, and thereby attempting to prevent or reduce problems associated with, a possible tendency of parents to be too demanding.

With regard to the claim that cloning even a single child would be psychologically harmful, it can be replied that such claims misuse the concept of 'harm.' Specifically, there is a serious problem with the claim that it would *harm* a child to bring her into being in circumstances where she would experience adverse psychological effects from being a clone. To explain this problem, we need to consider what it means to be harmed.

Harming versus Wronging

I shall draw upon Joel Feinberg's detailed and helpful discussions of what is involved in being harmed.[22–23] A key point is that persons are harmed only if they are caused to be worse off than they otherwise would have been. As Feinberg expresses it, one harms another only if the victim's personal interest is in a worse condition than it would have been had the perpetrator not acted as he did.[24] The claim that cloning harms the children who are brought into being, therefore, amounts to saying that the children are *worse off than they would have been if they had not been created.* Many readers will see problems with such an assertion. Some will say that it fails to make sense because it attempts to compare nonexistence with something that exists, and therefore it is neither true nor false. Others will maintain that it is *false.* Whether incoherent or false, it should be rejected. Because I have addressed the relative merits of these two criticisms elsewhere, I will not repeat that discussion, except to say that the better explanation of what is wrong with the statement seems to be that it is false.[25–26] To see its falsity, let us consider what a life would have to be like in order reasonably to say that it is worse for the person living it than nonexistence. I suggest that a life would have to be so filled with pain and suffering that these negative experiences greatly overshadow any pleasurable or other positive experiences the individual might have. If a neonate were born with a painful, debilitating, and fatal genetic disease, for example, we could reasonably make such a statement. However, the gap between such a neonate and a child cloned by an infertile couple is exceedingly great. Even if the cloned child experienced adverse psychological states associated with being a clone, that would not amount to a life filled with pain and suffering. Thus, the concept of harm is not appropriate for describing the cloned individual's condition.

I have applied the usual concept of harm to situations involving cloning. However, there is an objection that should be considered. It might be asserted that applying this concept of harm to this type of situation leads to implausible conclusions. Specifically, it seems to imply that it is ethically justifiable knowingly to create a child who will suffer disadvantages—even serious ones—as long as those disadvantages are not so severe that the life will be worse than nonexistence. To illustrate, consider a hypothetical situation in which a person with cystic fibrosis asks to be cloned, and a cure for cystic fibrosis is not yet available. Suppose that a physician knows about the cystic fibrosis but nevertheless provides the cloning, and the child later suffers the adverse effects of cystic fibrosis. It seems wrong for the physician to carry out the cloning in this example. According to the usual concept of harm, however, the child in this scenario is not harmed by being cloned, given that her life is better than nonexistence. If we cannot say that the child is harmed, then how can we account for our view that the cloning is unethical?

In reply, we can account for cloning being unethical in such a case without inventing a new concept of harm. Although the child is not harmed by being cloned, we can say that she is *wronged.* In particular, we can say that a certain right of the child is violated. It has been suggested that there is a type of birthright according to which people have a right to be born free of serious impediments to their well being.[27–30] It is important not to misunderstand this right; it is not a 'right to be born' but rather a right possessed by all persons who *are* born. Also, it is not what one might call a 'right against nature'; if a child is born with serious handicaps through the fault of no one, then the right in question is not violated. The right would only be violated if someone negligently or intentionally created a child with the requisite handicaps. I suggest that we can account for the wrongness of cloning in the cystic fibrosis example by positing such a birthright, a right that one's circumstances of birth be free of impediments that would seriously impair one's ability to develop in a healthy manner and to realize a normal potential. Negligently or intentionally creating a child who faces such severe impediments is a violation of that right, which I shall refer to as a right to a decent minimum opportunity for development. The handicaps imposed on the child with cystic fibrosis are not so severe that we could reasonably say that her life is worse than nonexistence, but they are severe enough to impair seriously her ability to develop.

An objection can be made against using the concept 'being wronged' to describe what is unethical about cloning in the cystic fibrosis example. Specifically, if the child were to claim that she was wronged by those who created her, it would commit her to the judgment that their duty had been to refrain from doing what they did; but if they had refrained, it would have led to her never being born, an even worse result from her point of view. Thus, it is argued, the child cannot reasonably claim that her creators should have acted otherwise, given that her life with cystic fibrosis is preferable to nonexistence. If she cannot make this claim, then she has no genuine grievance against them and cannot claim that they wronged her.[31–32]

However, this objection is mistaken. Perhaps we can see this more clearly by considering a similar type of situation. Instead of acts that cause a person to come into being, let us consider actions that cause an existing person to continue to live—that is, *life-saving* actions. Both types of acts have the result that a person exists who otherwise would not be in existence. Consider a patient who has suffered substantial bleeding because of a ruptured ulcer. Suppose that blood transfusions are necessary to save her life but she refuses transfusions on religious grounds and is considered mentally competent. Imagine that a physician provides transfusions despite her refusal, with the result that the patient's life is saved. Let us suppose, also, that the patient later states that she is better off having been kept alive. According to the objection in question, the patient cannot reasonably claim that the physician should have refrained from treating, given that her continued life is better than nonexistence. If she cannot make this claim, then she cannot claim that the physician wronged her.

But this conclusion is incorrect. Clearly, the patient can validly claim that she was wronged; her rights to informed consent and self-determination were violated. The benefits caused by the physician's act do not alter the fact that these rights were infringed. If a beneficial outcome removed all wrongness of the act, that would mean that paternalism, when successfully carried out, is ethically justifiable. This would be inconsistent with the view that competent patients have a right to refuse life-saving treatment. Similarly, the fact that the cloned child with cystic fibrosis has a life that is better than nonexistence does not mean that the child was not wronged. More generally, a child who is intentionally or negligently brought into being in circumstances where she lacks a decent minimum opportunity for development is wronged, even if her life is better than nonexistence, just as a competent patient who is forced to receive life-saving treatment is wronged, even though her subsequent life is preferable to nonexistence.[33]

Assuming there is a right to a decent minimum opportunity for development, the question arises concerning how serious the impediments must be in order for the right to be violated. No doubt, there is room for disagreement concerning this issue, and a sharp line probably cannot be drawn. Nevertheless, a basic concept can be stated: the impediments must be severe, not minor. As examples, I would suggest that creating a clone who has cystic fibrosis or Down's syndrome would violate the right in question, but creating a clone with, say, nearsightedness would not in itself constitute violation of the right.

Can we say, in the infertility case, that the cloned child would experience psychological problems of such magnitude that the right to a decent minimum opportunity for development would be violated? I suggest that the answer would depend largely on the approach taken by the parents in raising the child. Consider a hypothetical world in which the parents of cloned children always undermine their self-esteem. Then we might reasonably say that being a clone is associated with such serious impediments that the act of cloning violates the child's right to a decent minimum opportunity for development. Consider another hypothetical world in which many but not all parents of cloned children undermine their self-esteem. Then we might reasonably say that cloning puts a child at risk of experiencing obstacles severe enough to constitute a violation of the right in question. Depending on the level of risk involved, we might decide that cloning is wrong because the risk is unacceptably high. In yet another hypothetical world, the parents of cloned children are no more likely to undermine their self-esteem than the parents of other children. In that world it would not be reasonable to say that cloning in itself violates the right in question. If we were to allow infertile couples to use cloning, then which of these hypothetical worlds would the real world be most like? The fact is, we do not know. And this is the problem with the argument that cloned children would experience severe psychological obstacles to well-being; it is based on empirical assumptions concerning how cloned children would generally be treated by parents and others, and we lack evidence supporting those assumptions.

The Argument from Parent-Child Relationships

Let us consider the argument that children would be objectified and parent-child relationships generally would be adversely affected. This argument arises from reflection on what it would be like if there were a widespread practice of controlling the characteristics of our offspring. This practice might involve the insertion and deletion of genes in human preembryos, as well as cloning and other laboratory techniques not yet envisioned. In some cases, such manipulations might have a therapeutic goal; perhaps disease-causing genes would be replaced by normal ones. Objections to such therapeutic manipulations have been based mainly on concerns about whether they can be performed safely. But other forms of genetic control might have the much different goal of enhancing offspring nondisease characteristics, such as height, intelligence, and body build, and it is especially this type of control that raises concerns about undesirable changes in the attitudes and expectations of parents toward their children.[34–35] The specific concerns can be expressed by a number of questions: If a child failed to manifest the qualities she was designed to have, would the parents be less inclined to accept the child's weaknesses? Would children be regarded more as objects and less as persons? Would less tolerance for imperfection result in less compassion toward the handicapped? Would children who recognize their own shortcomings blame their parents for failing to design them better? Would such feelings sometimes disrupt family relationships? Would knowledge of being designed make a child feel more controlled by parents? Would this result, for example, in greater adolescent rebelliousness? These and other questions suggest a number of ways in which disharmony could enter into parent-child relationships.

However, a reply can be made. Although these are important concerns, their bearing on infertility cases is tangential. Because cloning does not involve inserting and deleting genes, concerns over whether these particular manipulations can be done safely are not directly applicable. Also, cloning in the infertility cases does not involve efforts to enhance the child's genetic makeup. Thus, the concerns expressed above that are specific to enhancement do not directly apply, either; these include the concern that modifying children in order to enhance their characteristics objectifies them. The claim that cloning in the infertility case objectifies the child is weakened by the fact that the purpose is not to design the child but to have a genetically related child, in the only way that is possible. In addition, the number of cases like the one being considered—those in which there is both male and female infertility and use of family donors has been ruled out—would be relatively small. Thus, it is difficult to argue that cloning restricted to such cases would result in widespread changes in parent-child relationships. For these reasons, the arguments in question do not succeed in showing that cloning in the infertility cases would be wrong.

The Argument Based on Abuses

The third argument is that abuses might occur if the technology of cloning is used by unscrupulous persons. A paradigm of such envisioned abuse is found in Aldous Huxley's *Brave New World,* in which multiple genetically identical persons are produced and conditioned to fill defined social roles.[36] Other variations of such abuse could be imagined, in which cloning plays a role in the systematic control of persons by determining their genetic makeup and upbringing.

In reply, it seems clear that we can distinguish such abuse from the infertility case in question; the couple's trying to have a child would not constitute such abuse or even remotely approach it. Perhaps it will be objected that permitting cloning in the infertility cases would facilitate development of the technology, thereby making it more likely that unscrupulous persons could use it. In reply, this objection assumes that human preembryos would not be created by cloning as part of research. In the future it might become useful to create such preembryos, not for the purpose of transferring them to a woman's uterus, but for studies in any of a number of scientific areas, perhaps including preembryo development, cell differentiation, immunologic properties of cells, or the creation of cell lines using stem cells. Some of these scientific areas—or others not mentioned here—might be considered important enough in the future to justify creating human preembryos by cloning for purposes of laboratory studies.[37] Thus, the technology of cloning humans might go forward, even if use of that technology to produce babies is proscribed.

Although the main arguments against cloning are not persuasive when applied to the infertility cases, at least two additional arguments can be presented that focus more directly on those cases. First, consider a son who is, as we might put it, genetically identical to his mother's husband. Some might be concerned by the somewhat oedipal nature of this situation. Would knowing that one's mother is sexually attracted to someone genetically identical to oneself cause special psychological problems for the child? Would the ill effects be great enough to make life worse than nonexistence? Would the right to a decent minimum opportunity for development be violated? Second, consider a father whose daughter is genetically identical to his wife. Would there tend to be a higher incidence of sexual abuse in this type of situation? These questions raise legitimate concerns, but because the answers are highly speculative at present, these concerns do not constitute definitive arguments against the particular use of cloning in question. However, they suggest possible topics for inclusion in preimplantation counseling, if such counseling is provided.

In summary, some main reasons have been identified supporting freedom to use cloning in the type of infertility case being considered. Also, each of the main arguments against cloning has been shown to involve substantial difficulties when applied to the infertility cases. It seems reasonable to conclude that the arguments against cloning are not compelling enough, either singly or collectively, to support the conclusion that cloning is wrong in such cases.

Cloning versus Collaborative Reproduction

It might be objected that it would be ethically preferable for the couple to have children who are not genetically related to them. Adoption would be a possibility, or donor gametes or preembryos could be used. If the woman is capable of gestating using donor gametes or preembryos, then at least she could be biologically related to the child. In reply, some infertile couples prefer not to adopt, even if that is the only way they can have children to raise. Moreover, if the couple tried to adopt, there would be a significant chance that their attempt would be unsuccessful because of the difficulties involved in adoption.

The claim that using donated gametes or preembryos would be ethically preferable to cloning overlooks the problems associated with third-party collaborative reproduction. Such arrangements raise a number of difficult issues because of the separation of genetic and social parenthood: What should the children be told about their origins? When and how should any informing take place? What if the child later wants to meet the genetic parents?

For example, when there is male-factor infertility, the man often prefers secrecy concerning his inability to beget. As a result, couples often choose not to reveal the fact of gamete donation to the child and others. This creates the problem of there being a significant deception at the center of the family relationship. Maintaining this deception can take its toll, including a substantial emotional burden on the couple. Also, such deception is at odds with the values of honesty and trust that should bind families together. On the other hand, children who are told the truth about their origins might develop strong desires to meet their genetic parents. If these wishes are frustrated, the result might be substantial emotional distress for the child. Thus, depending on how these various issues are handled, adverse psychological consequences are possible for the child and family. It would be reasonable for the couple to prefer not to encounter these problems, and cloning would provide a way to avoid them. The existence of these problems calls into question the claim that third-party collaborative reproduction would be less ethically problematic for the couple than cloning.

It might be objected that cloning would also raise difficult issues for the couple. For example, should the fact of cloning be revealed to the child, and if so, when and how? If the child is not told, then she will not suffer any psychological ill effects arising from knowledge of being a clone. But if such children were never told, then they would be deprived of important information about their background. Should family and friends be told about the cloning? Or should they be led to believe that the child simply "looks like" one of the parents?[38] There might be disagreements between husband and wife over whom, when, and how to tell. In reply, although these too are difficult issues, their existence seems to indicate that the two approaches to reproduction are at a standoff in this regard; both raise issues that carry the potential for interpersonal conflict within the family. Because of this parity, it is not obvious that third-party collaborative reproduction is ethically preferable to cloning.

I have been discussing a situation in which cloning is the only way for a couple to have a child genetically related to one of them. However, the arguments I have stated in support of cloning are applicable to other infertility cases in which the alternative is third-party collaboration. Suppose the woman cannot produce ova, but donor ova and the husband's sperm could be used. The couple might nevertheless prefer cloning in order to avoid the complications associated with third-party gamete donation discussed above.[39] Alternatively, suppose the man cannot produce sperm but the woman's ova could be used with donor insemination. A preference for cloning might be based, not solely on male ego, but also on a desire to avoid the problems associated with third-party reproduction. Moreover, the three main arguments against cloning do not fare better when applied to these types of infertility cases. Again, cloning does not harm the child, nor is it clear that the right to a decent minimum opportunity for development would be violated. The argument that parent-child relationships generally would be adversely affected continues to be unpersuasive because enhancement is not involved and, although we now are dealing with a larger class of infertile couples, the number still is relatively small compared to the general population. These scenarios also would be distinguishable from *Brave New World* abuses. Thus, the arguments against cloning do not constitute compelling reasons to override the freedom of infertile couples to use cloning in these cases either.

Cloning as a Bridge to Future Remedies for Infertility

Research in gene therapy is resulting in the discovery of ways to insert genes into human cells. This is increasing the plausibility of the view that in the future it might be possible to insert, and perhaps delete, a variety of chosen genes. Such modifications could be performed on cells prior to using them for cloning. By means of such techniques, a child created with cloning technology could have genes from both parents. Starting with a cell from one parent, one might change hair color, skin complexion, and eye color, using genes from the other parent. Perhaps genetic defects causing infertility and susceptibilities to other diseases would also be corrected. The child then would be genetically unique. Thus, the argument against cloning based on lack of genetic uniqueness would no longer be applicable. Prohibiting cloning in the future might prevent us from helping infertile couples in these ways.

Such modifications would not necessarily include changes that constitute 'enhancement,' such as higher intelligence, better body build, or greater height; the goal could be to make the child genetically different from either parent, rather than to produce a 'superior' child. In that event, objections to genetically enhancing our offspring would not be applicable. Moreover, this particular use of genetic technology—to help an infertile couple have a child—would also be distinguishable from *Brave New World* abuses.

This type of reproduction would not be cloning, strictly speaking. The term "cloning" in both scientific and lay usage, implies the production of a genetically identical copy. In fact, we lack a common term to refer specifically to the type of reproduction being envisioned.

It might be objected that we could not ethically create children in this manner because developing the technology would involve experimenting on unconsenting subjects. It might be claimed, for example, that it would be unethical to try to alter genetically a child's hair color because some unintended adverse genetic modification might occur. It might be argued that altered hair color is not a significant enough benefit to justify the risks involved. In reply, it is conceivable that our technology might advance to the point where the risks involved in such an attempt would be low. Moreover, if being a clone exposes one to the risk of adverse psychological effects, as opponents of cloning maintain, then alterations that prevent one from being a clone might have benefits significant enough to outweigh the risks. Therefore, I do not believe we can reasonably claim that such genetic modifications could never ethically be done. If such modifications could be performed safely, this type of reproduction might be ethically *preferable* to reproduction using donor gametes or preembryos because the problems associated with third-party collaboration would be avoided.

Should Exceptional Cases Be Permitted?

I have argued that cloning humans could sometimes be ethically defensible in cases of infertility. It might be objected that cloning should not be permitted even in those cases in which it is ethically justifiable. One argument supporting this objection begins by claiming that a general prohibition of human cloning is warranted, based on the reasons against cloning discussed above. Legal restrictions are needed to prevent the creation of cohorts of multiple clones, as well as other clear abuses of cloning technology. Restrictions also are needed to prevent a widespread practice of cloning, thereby avoiding the feared ill effects on parent-child relationships. Moreover, it is claimed that practical problems involved in attempting to enforce a general policy against human cloning while permitting exceptions provide grounds for not allowing the exceptions.[40] In particular, it would be difficult for authorities to gather the evidence needed to distinguish allowable from nonallowable cases. For example, fertile couples might have children by cloning, yet claim that they are infertile. Prosecution could not reasonably proceed unless evidence ruling out infertility were obtained. Assuming that an infertility doctor assisted the couple in the cloning, often that doctor's testimony and records would be crucial evidence. However, such confidential records are protected in the absence of a court order, and establishing 'reasonable cause' for such court orders might be difficult in many cases. Moreover, the couple could refuse to release the records voluntarily, claiming (perhaps disingenuously) that the information is too personal and sensitive. The legal protection of medical records behind which such couples could hide is itself important and based on constitutional guarantees against unreasonable invasions of privacy. It can be argued that relaxing such protections in order to distinguish permitted from nonpermitted cloning would be too intrusive of reproductive privacy. Thus, if we permit cloning in the infertility cases, in effect we open the door for anyone to use cloning and get away with it.

In reply, it is possible to have widespread compliance with a law even though there are difficulties in detecting violations. If we were to make cloning, except in infertility cases, illegal not only for the couples using it but also for the physicians carrying it out, I believe that we would see widespread compliance. Most physicians will choose to avoid illegal activity, even if authorities would face difficulties in detecting violations. This tendency could be reinforced if the penalties for being convicted of the violation are high, even though the likelihood of detection is relatively low. Although a few cases of cloning might occur outside the allowed exceptions, it is doubtful that we would see the sort of widespread practice that the policy in question would attempt to prevent. Reproductive privacy can be protected while having a generally effective policy that prohibits cloning except in cases of infertility.

Another objection is that it would be inconsistent to permit cloning for infertile couples but not fertile ones. If reproductive freedom is important enough to permit infertile couples to use cloning, then why shouldn't all couples be allowed to use it, if that is their desire? According to this argument, if permitting the infertile to clone children commits us logically to allowing everyone to do so, then we should not allow the infertile to clone.

A reply to this objection can be based on the fundamental reasons we give in explaining why procreative freedom is worthy of protection. I identified six reasons that can be given in helping explain why freedom to have children is important. It is worth noting that the goals and values reflected in those six reasons can be achieved by fertile couples without resorting to cloning. By having a child through sexual intercourse, they can: participate in the creation of a person; affirm their mutual love through procreation; deepen their sexual intimacy; obtain a link to future persons; and have experiences associated with pregnancy, childbirth, and child rearing. Therefore, in prohibiting those who are fertile from using cloning, we do not deprive them of the ability to pursue what is valuable about having children. However, when we forbid the infertile to use cloning, we force them to choose either not to realize any of those valued goals or to pursue collaborative reproduction with its associated difficulties. This is the morally relevant difference, I would suggest, that can justify differing policies for fertile and infertile couples. In conclusion, there do not appear to be good reasons to disallow exceptions to cloning for infertile couples.

Notes

1. In this paper, the term "cloning" refers to creating a child by transferring the nucleus of a somatic cell into an enucleated egg cell. This method should be distinguished from blastomere separation, which involves the division of a preembryo when its cells are totipotent. Although both produce individuals with identical chromosomes, the two methods have different ethical implications. For example, cloning by nuclear transfer involves the possibility of creating numerous duplicates of the original individual, but in blastomere separation only a few copies can be produced, as explained in Cohen J, Tomkin G. The science, fiction, and reality of embryo cloning. *Kennedy Institute of Ethics Journal* 1994; 4:193–203.
2. National Bioethics Advisory Commission. *Cloning Human Beings: Report and Recommendations of the National Bioethics Advisory Commission.* Rockville, Maryland: National Bioethics Advisory Commission, 1997:79–81. World Wide Web: http://www.nih.gov/nbac/nbac.htm.
3. Winston R. The promise of cloning for human medicine: not a moral threat but an exciting challenge. *British Medical Journal* 1997; 314:913–4.
4. See note 2, National Bioethics Advisory Commission 1997: ii–iii,13,23–4.
5. The researchers who produced Dolly used nuclei from three sources: late embryos, fetal cell cultures, and cell cultures derived from the mammary gland of an adult sheep. Of 277 preembryos created using mammary cells, only one developed into a live lamb. Sixty-two percent of fetuses from all three sources failed to survive until birth, compared to an estimated 6% fetal loss rate after natural mating. This high rate of fetal loss suggests an increased incidence of genetic anomalies. For data on the total number of preembryos and live births, see Wilmut I, Shnieke AE, McWhir J, Kind AJ, Campbell KHS. Viable offspring derived from fetal and adult mammalian cells. *Nature* 1997; 385:810–3.
6. Stewart C. Nuclear transplantation: an udder way of making lambs. *Nature* 1997; 385:769,771.
7. This situation could result from various medical conditions: the woman's ovaries might have been surgically removed, or she might suffer from premature ovarian failure—the inability of ovaries to produce ova; the man could have testes that produce no sperm, or perhaps a small number of sperm are produced but attempts to perform intracytoplasmic sperm injection (ICSI) using donor ova have been unsuccessful, among other possibilities. For a discussion of the uses and limitations of ICSI, see Silber SJ. What forms of male infertility are there left to cure? *Human Reproduction* 1995; 10: 03–4.
8. Recer P. Clone fear may slow research. *AP Online* 1997; Mar 5:19:20EST.
9. Nash JM. The age of cloning. *Time* 1997; Mar 10: 60–61,64–5.
10. Butler D, Wadman M. Calls for cloning ban sell science short. *Nature* 1997; 386:8–9.
11. Masood E. Cloning technique "reveals legal loophole." *Nature* 1997; Feb 27. World Wide Web: http://www.nature.com.

12. Harris J, Is cloning an attack on human dignity? *Nature* 1997; 387:754.
13. Pohlman E, assisted by Pohlman JM. *The Psychology of Birth Planning.* Cambridge, Massachusetts: Shenkman, 1969:48–81.
14. Pohlman E. Motivations in wanting conceptions. In: Peck E, Senderowitz J, eds. *Pronatalism: The Myth of Mom and Apple Pie.* New York: Crowell, 1974:159–90.
15. Veevers JE. The social meanings of parenthood. *Psychiatry* 1973;36:291–310.
16. Arnold F. *The Value of Children: A Cross-National Study.* Honolulu: East-West Population Institute, 1975.
17. Laucks EC. *The Meaning of Children: Attitudes and Opinions of a Selected Group of U.S. University Graduates.* Boulder, Colorado: Westview, 1981.
18. Gould RE. The wrong reasons to have children. In: Peck E, Senderowitz J, eds, *Pronatalism: The Myth of Mom and Apple Pie.* New York: Crowell, 1974:193–8.
19. Kahn A. Clone mammals . . . clone man? *Nature* 1997; 386:119. Kahn states, "the debate has in the past perhaps paid insufficient attention to the current strong social trend towards a fanatical desire for individuals not simply to have children but to ensure that these children also carry their genes."
20. Strong C. *Ethics in Reproductive and Perinatal Medicine: A New Framework.* New Haven: Yale University Press, 1997:13–22.
21. Ellin J. Sterilization, privacy, and the value of reproduction. In: Davis JW, Hoffmaster B, Shorter S, eds. *Contemporary Issues in Biomedical Ethics.* Clifton, New Jersey: Humana Press, 1978:109–25,
22. Feinberg J. *Harm to Others.* New York: Oxford University Press, 1984:31–64.
23. Feinberg J. Wrongful life and the counterfactual element in harming. *Social Philosophy and Policy* 1987; 4:145–78.
24. See note 23, Feinberg 1987:149. Feinberg's discussion is more extensive; this is only one of six conditions he identifies as necessary and sufficient for harming, pp. 150–53.
25. See note 20. Strong 1997:90–92.
26. Also see note 23. Feinberg 1987:158–9.
27. Bayles MD. Harm to the unconceived. *Philosophy and Public Affairs* 1976; 5:292–304.
28. Also see note 22, Feinberg 1984:99.
29. Steinbock B, McClammock R. When is birth unfair to the child? *Hastings Center Report* 1994;24(6):15–21.
30. Also see note 20, Strong 1997:92–5.
31. This type of objection is put forward by Feinberg. See note 23, Feinberg 1987:168.
32. A similar objection is stated by Brock DW. The non-identity problem and genetic harms—the case of wrongful handicaps. *Bioethics* 1995; 9:269–75.
33. For this response to the objection in question see note 20. Strong 1997:93–4.
34. Botkin JR. Prenatal screening: professional standards and the limits of parental choice. *Obstetrics and Gynecology* 1990; 75:875–80.
35. Strong C. Tomorrow's prenatal genetic testing: should we test for 'minor' diseases? *Archives of Family Medicine* 1993; 2:1187–93.
36. The method of genetic duplication that Huxley described did not involve replacement of the nucleus of an egg cell. In his fictional account, the fertilized egg was described as "budding" when "Bokanovsky's Process" was applied to it; the result could be as many as "ninety-six identical twins."
37. See note 3, Winston 1997.
38. We can imagine friends saying, for example, "He's the spitting image of his father," but not realizing, because of the age difference, that they are genetically identical.
39. Cloning in such a case would involve an ovum donor, but the chromosomes would be removed from the ovum. Although mitochondrial DNA in the ovum would be inherited by the child, the ovum donor would not be a "genetic mother" in the ordinary sense of that term. Thus, although there would be third-party collaboration, it would not involve the difficulties typically associated with third-party genetic parentage.
40. This type of objection is suggested in *Cloning Human Beings.* See note 2, National Bioethics Advisory Commission 1997:81–2.

Four Questions about Ethics

by Gregory E. Pence

Gregory E. Pence is Professor of Philosophy at the University of Alabama in Birmingham.

> *The first stage [of modern moral philosophy] is one of gradual emergence from the traditional assumption that morality must come from some authoritative source outside of human nature, into the belief that morality might arise from resources within human nature itself. It was a movement from the view that morality must be imposed on human beings towards the belief that morality could be understood as human self governance or autonomy. This stage begins with the Essays of Michel de Montaigne and culminates in the work of Kant, Reid, and Bentham.*
>
> *During the second stage, moral philosophy was largely preoccupied with the elaboration and defense of the view that we are individually self-governing, and with new objections and alternatives to it. The period extends from the assimilation of the works of Reid, Bentham, and Kant to the last third of the present century.*
>
> *[In the last stage today], the attention of moral philosophies has begun to shift away from the problem of the autonomous individual toward new issues concerning public morality.*
>
> —J.B. SCHNEEWIND, "MODERN MORAL PHILOSOPHY"[1]

IN THIS CHAPTER, I describe four questions to ask when thinking about the morality of human asexual reproduction. Before these descriptions, it will be helpful to have a case for focus. (This case, although realistic, does not refer to an actual case.)

CASE #1—SARAH AND ABE SHAPIRO

Sarah and Abe Shapiro yearned for a child for years before being able to have one. Both came from large Jewish families that put great emphasis on parental involvement with children and on family activities such as playing sports, eating nightly meals, and going on long camping trips.

Sarah and Abe also inherited something else from their families. Tay-Sachs disease runs in Jewish families of Eastern European origin. It is a lethal genetic disease that produces children who always die before they become teenagers.

Knowing their risk, Sarah and Abe used in vitro fertilization (IVF) so that any embryo implanted in Sarah could be screened for Tay-Sachs. In IVF, three embryos are often implanted in hopes that one will successfully gestate.

Such was the way Michael was created. Unfortunately, when the embryo that would become Michael was moving down Sarah's fallopian tube, it damaged her tube and rendered her infertile (her other tube was already damaged). So the Shapiros resigned themselves to having one child of their own and hoped, perhaps, to adopt another later.

When Michael was four he and Abe were driving home from an outing when a drunk driver smashed into their car, instantly killing Abe and rendering Michael comatose, but with a beating heart sustained on a respirator.

After Abe's funeral, Sarah hoped for Michael to recover, praying to God for a miracle which unfortunately did not come. During this time, she mourns the death of both Michael and Abe. After a year, her rabbi and therapist urged her "to move on with your life." They want her to agree to remove the respirator and allow Michael's body to die. She is only 40.

Sarah does not want to remarry. She is a writer and now owns her own home because of Abe's life insurance. However, she misses having a child in the house.

At this point, she decides to have one of her eggs removed, its nucleus taken out, and have the genes from Michael's body inserted in her egg to create a new embryo. After doing so, she will let Michael go. One of her reasons for using Michael's genes, she says, is that, "I couldn't bear to have a child who then died very young of Tay-Sachs." In this way, she knows her child will also be normal and be part of both her and Abe.

In many sessions, the rabbi, therapist, and infertility-physicians explore with Sarah the idea that she is merely attempting to replace Michael and that she has not fully accepted Michael's death. These professionals want to ensure that Sarah understands that the new child will be very different from Michael. They emphasize that Abe's influence will be missing, that Sarah's egg will contribute mitochondrial genes, that Sarah herself is now different, and so on.

Sarah claims that she is not trying to mechanically replace Michael and that she has accepted Michael's real death. She adds, "I know Michael and Abe are dead, but if God lets me bring forth this new child, whom I will call David, then Michael's and Abe's lives will not have been for nothing, for in David's life I can see, if not them, then at least their features and talents live on. Maybe I'll see Michael's way of laughing and Abe's swagger after he performs well in sports. What's wrong with that?"

A genetic counselor points out that she may also get the worst qualities from Abe and Michael, and is she prepared for that? "The worst qualities?" she ponders. "Well, they sure weren't perfect and they did have some of those, but I personally would rather have their worst qualities than just accept some anonymous sperm implanted in me, where the child will have no relation to Michael or Abe, and perhaps, to a history of Jews going back five thousand years."

Query 1—Does the Rule Intrude Too Much on Personal Liberty?

John Stuart Mill wrote *On Liberty* in 1859, and it contains an admirable distinction between private life and public morality, a distinction based on the concept of harm. Mill believed that a civilized society must promote certain ideals and discourage certain vices. Society can do this through its public policy while granting individuals a sphere of private action that is protected from interference by government. Power of the nation-state can be dangerous when used against the individual, and so the agents of government—such as police and military—should be forbidden to meddle in private life.

Equally, Mill held, the majority of citizens should be forbidden from becoming tyrannical. It should be forbidden from imposing its religious beliefs on a dissenting minority, even indirectly—say, by a judge who insists on a Christian prayer with a jury before they hear a case. It should be forbidden from censuring what is discussed publicly, say by a television station that decides that its viewers should not see homosexual characters. It is important to emphasize here that Mill believed that the majority's tyranny is normally done in the name of morality.

It is natural to ask where the line is to be drawn between private and public life. Mill's rough rule-of-thumb is called his *harm principle*: private life encompasses those actions of adults that are purely personal and put other people at no risk of harm. In such areas, there should be no interference by government—even for a person's own good. Consider non-traditional sex roles between two consenting adults (where the wife leaves home to work and the husband raises the children at home): even if other people consider these roles immoral, their relationship for Mill poses no moral question because no one else is affected.

Building on Mill's work, we can distinguish between four different areas where issues about human cloning arise: (1) personal life, (2) morality, (3) public policy, and (4) the law. Issues of personal life are purely private and affect no one else. When someone else is affected, issues move from the personal to the realm of morality. When society attempts to promote certain positive values while at the same time tolerating personal disagreement with those values, we move into the third area, public policy.

Actions in the area of public policy, like those in the area of morality, do affect other people's interests, but persuasive actions in public policy are not necessarily condemned as immoral. Consider alcohol. Although society tries to discourage consumption of alcohol (by taxation) and regulates it (forbidding alcohol at elementary schools) people may drink in their homes without being viewed as immoral. Consider also adoption. Society wants adults to adopt needy children, and offers tax incentives to adults to so encourage this, but no one thinks it immoral for a childless couple not to adopt a baby.

These spheres overlap and shade into each other, and there are no exact criteria for separating one area from the next. The general goal is to limit the range of morality from two ends: first by carving out a zone of private, personal life, and second, by allowing society to encourage and discourage behaviors in public policy without explicit moral judgment or legal penalty. The general goal recognizes that we are all better off not moralizing every aspect of life.

One of the things Mill meant is that views that are essentially religious, even if held by the majority, should not be imposed on the minority. Especially in areas so personal as the make-up of the family and familial reproduction, the religious views of the majority have no place running federal policy.

Query 2—What Is the Point of the Moral Rule?

Instead of the usual question about ideal morality (about how morality ought to be), it is useful to consider how morality actually works. Call this the functional view of morality.

In this view, moral rules exist to adjudicate conflicts between the interests of persons. Because modern society contains many different kinds of people with many different points of view, moral rules are necessary for us to get along peacefully. In this functional view, the point of moral rules is not to prepare everyone for salvation or to create a purely religious state on earth. (These were the metaphysical beliefs associated with moral rules that at one time were quite functional.) Nor is the point of morality to create the greatest good for the greatest number of humans and animals on the planet. Nor is the point to create a perfectly rational, elegant theory of morality. Instead, the point is the more minimal one of getting along in a world where some resources will always be scarce, where interests of people conflict, and where people are interdependent and must cooperate.

So moral rules adjudicate social relations. Where they fail, the tougher ways of the law begin. Given that function, past moral rules may not always work in contemporary times, and when that happens, the nature of morality itself comes into question. For example, the very concept of having an interest has changed substantially over the last century, from covering one's household property to covering one's interest in pirated copies of one's book sold in China.

Moral rules in this functional sense are moot when there is no conflict, where the people have no real interests at stake, or where there are no existing people. For example, suppose Smith and I agree to share the planting of a boundary hedge along the property line on the west side of my yard, but my neighbor Jones on the eastern side is jealous of the cooperation between Smith and me, so much so that he objects to the joint project between Smith and me. Jones has no right to do so because he has no interests at stake. As so often happens in morality, his very objection creates a moral issue between Jones and me (because he is trying to interfere with my relations with my other neighbor) when there was no moral issue before.

Thus the point of moral rules is not to create an ideal society. Some philosophical vision of the future must do that, while moral rules allow us to get along enough to get there. In the technical language of moral philosophy, there is the theory of the right and the theory of the good. If we have the right theory of the right, we will allow different people to live their lives according to their view of the good.

Application of this point to human asexual reproduction is obvious: if there is no conflict between two or more people, there is no moral issue present. Despite the widespread belief to the contrary, if no one is harmed by human asexual reproduction, then it raises no moral issue.

I want to also make a more general point here about the point of moral rules. The two great traditions that we have inherited from the past focus on two ways of evaluating moral acts: by their motives and by their consequences. Hence, if we want to know why an action is right, we can look at either the motives of the agents or the action's consequences. Judaeo-Christian ethics tends to focus on the motive of the act—what was in the agent's mind or heart—and not on what consequences occur. A secular ethics such as utilitarianism focuses on the actual consequences.

This surprisingly simple fact—that motives and consequences determine the morality of an act—is a helpful one to keep in mind when we ask why a certain rule is still a good one. Sometimes, a rule will become written in stone and we forget why it came about in the first place. If we carefully inspect the motives and consequences associated with that rule, we may sometimes discover that it is outdated.

Nor should we assume that the specific moral judgments that we make and that seem "obvious" to us will stand the test of time. As the Australian moral philosopher Peter Singer writes on this question:

> Why should we not rather make the opposite assumption, that all the particular moral judgments we intuitively make are likely to derive from discarded religious systems, from warped views of sex and bodily functions, or from customs necessary for the survival of the group in social and economic circumstances that now lie in the distant past? In which case, it would be best to forget all about our particular moral judgments and start again from as near as we can get to self-evident moral axioms.[2]

For example, our culture traditionally has forbidden actively assisting a terminally ill person to die ("active euthanasia"), but it considers it permissible to merely watch such a person die slowly. So a basic rule in our culture is that allowing terminal patients to die is permissible, where assisting them is not.

This rule is outdated. How do we know? Because in both modes, the motives and consequences are the same. Situations often arise with terminal patients where the motives of everyone—including the patient himself—are to create a quick, painless death. Here the people intend quick death and quick death is the result. Given such a situation, it cannot matter morally whether the actions taken to hasten death are passive or active.[3]

Put differently, if the motives were bad and the consequences were bad, then the action would be bad according to either kind of moral theory. But it would make little sense to say that the description of the act as passive or as active really held the moral weight. To say so would be like saying that a performance of a piece of music was bad not because of how it was played but because of how it was classified.

Query 3: Why Assume the Worst Motives?

The case of Sarah Shapiro is deliberately formulated to have a parent who has good motives about originating a child asexually. As the case shows, such a possibility is not unimaginable and, given the unpredictability of human life, a case such as this will one day arise.

Most popular discussions about cloning a human assume the worst possible motives in parents, but why on earth make such assumptions? Without evidence? If someone assumes that every person he meets is a secret racist or anti-Semite, we say he is paranoid, or a misanthrope, or warped. Why assume the worst motives when we are thinking about morality? Or in public policy? This way of thinking got us nowhere in the cold war, when the U.S. and Russia competed in the nuclear arms race and where it was assumed that Communists were evil people and Americans were saints. Why assume in public policy what we don't assume in ordinary life? We don't forsake participation in car pools that take our kids places because we fear that some parent may decide to kidnap the kids for ransom. Why should we assume worse when it comes to thinking about parents in public policy?

It has always been a trick of advocates of the status quo to assume the worst motives in humans. That is what the theory of original sin is all about. But humans are a lot better off today than a thousand years ago, and also a lot better off than a hundred years ago. And the main reason why is the electricity, antibiotics, clean water, efficient transportation, mass communication, and public education that humans have created. (Those who disagree know only the false, rose-colored versions of history seen in the mass media.) So why not trust humans rather than fear them? Who else has brought this progress? (If God has allowed humans to progress, why won't he allow them to progress more?)

An important corollary here is to ask about the evidence for assuming bad motives in ordinary people. If there is no such evidence, no such motives should be assumed. We have thousands of years of history with human parents and we know them well. We know that most parents most of the time do not have evil motives toward their children.

Nevertheless, many of our pundits assume the worst about us. Catholic University law professor Robert Destro wondered if cloned humans would have adequate legal rights "if they were created to perform specific work."[4] Why assume this? It is like saying that we should not admit emigrants to this country because they might by enslaved by natives. Why would a parent be so bigoted? ("Laura, dear, why don't we clone a little slave-child to walk the dog and clean the kitty litter?")

The Reverend Richard McCormick said that "the obvious motives for cloning a human were 'the very reasons you should not.' "[5] Obviously, Father McCormick thinks it is "obvious" that couples have bad motives. He thinks that a couple might try to "create someone who could be a compatible organ donor." Really? Create your son and rip out his heart?

McCormick was probably thinking of the Ayala case where a couple conceived a daughter as a possible donor of bone marrow for their elder daughter dying of leukemia, and where they were lucky and had a new baby whose marrow matched.[6] But as medical sociologist Jay Hughes notes, there is all the difference in the world between renewable resources for transplantation, such as bone marrow, skin, urine, hair, and blood, and non-renewable human resources, such as hearts.[7]

Bioethicist Thomas Murray, a member of the Bioethics Advisory Commission, said, "Why are we uneasy about cloning? We might be worried over the dangers of excessive control over human reproduction, about the dangers of unbounded human pride."[8] But why assume that a government ban on human cloning is also not "excessive control over human reproduction?" Why assume that "unbounded human pride" is why couples would originate children by cloning? Why is giving couples more control over baby-making—which they have lacked through 99.9% of human history—a bad thing?

Why make such ridiculous assumptions about the motives of ordinary couples yet to have children? Go to your local neighborhood meeting, Parents-Teacher Association night, or Kiwanis Club and ask yourself: are all those people the kind of people who have bad motives? To assume bad motives in a crack addict or an alcoholic parent is understandable because we know that their free will has been largely overtaken by a drug. The drug will win out over any motive for a child's welfare. But most parents are not drug-dependent, nor are they malignant narcissists. Indeed, when we are almost exclusively discussing parents who want and plan for a child, and have good resources to raise such a child, we have adverse selection into that subset of parents who are unlikely to have such bad motives.

Query 4—Why Fear Slippery Slopes?

One of the central objections to cloning a human concerns the idea of a slippery slope, perhaps the second most famous idea in ethics (behind the Golden Rule). True, it will be allowed, extraordinary circumstances may make it plausible in the Shapiro case to think about allowing human asexual reproduction, but if that case is allowed, then another similar case must be allowed, until we get to some really terrible scenarios.

For example, twenty years ago in the debate about in vitro fertilization, Leon Kass objected that:

> At least one good humanitarian reason can be found to justify each step. The first step serves as a precedent for the second and the second for the third, not just technologically but also in moral argument. Perhaps a wise society would say to infertile couples: "We understand your sorrow, but it might be better not to go ahead and do this."[9]

The rough idea here is that if a small, benign change is allowed, it will inevitably lead to another, less benign change, and so on through a series of inevitable steps, until a point is reached where a very bad outcome is at hand. A corollary is that, once the first change is accepted, there is no easy way to stop until the last, bad point is reached. Hence, the inference is made, better not to change at all.

The slippery slope is, for better or worse, also a central idea in bioethics. Because bioethics has been at the forefront of change over the last decades, "slope predictions" have been common. Indeed, every time real social change occurs, it scares most people, and some moralists will predict that the sky will soon fall: "The dawn of the era of cloning is a little like splitting the atom," said Dr. Glenn

Bucher, president of the Graduate Theological Union in Berkeley, California, "with enormous prospects for evil and enormous prospects for good."[10]

But we must not be manipulated by predictions made at the drop of a hat. In the one above, with what is Bucher comparing "enormous prospects for evil"? The Holocaust? The Mongol invasion of Europe? AIDS? Does he really mean to indirectly refer to the atomic bomb?

One famous book was full of slope predictions. Thirty years ago, Alvin Toffler breathlessly coined the term "future shock" to "describe the shattering stress and disorientation that we induce in individuals by subjecting them to too much change in too short a time."[11] His *Future Shock* sold millions of copies and he was anointed as the futurologist whose omniscience revealed the (mostly dire) future of humanity. Toffler hyperventilated that social change was occurring so fast that we were losing all our moorings and would soon be adrift in a sea of social chaos. (Alasdair McIntyre's books push the same theme at the theoretical level in ethics.[12])

Toffler wrote *Future Shock* between the years of 1965 and 1970, when the industrialized, Western world was rapidly changing. Those years witnessed big changes in music, sex roles, blended families, suspicion of authority and old age, and a new tolerance for drugs, sexual experimentation, contraception, abortion, and divorce.

What Toffler failed to predict was that too much change creates an opposing reaction toward stability. By 1981, when AIDS began, the conservative reaction was already well under way and it kept rolling through the 1990s: couples reverted to traditional sex roles, nuclear families were again seen as an ideal, hostility renewed towards illegal drugs (especially cocaine and heroin), realization occurred that contraception and abortion weren't stopping teenage pregnancy, and divorce was seen to hurt children and hence, to be too easy. If we slipped down the slope, and many would deny we did, then at some point we took stock of where we were, changed our minds, and walked back up.

The specific predictions made by *Future Shock* about human cloning, artificial wombs, and genetic engineering are lessons in caution. Nobel Laureate geneticist Joshua Lederberg predicted to Toffler—some time between 1965 and 1970—that "somebody may be doing it [cloning] right now with mammals. It wouldn't surprise me if it comes out any day now."[13] As for cloning humans, Lederberg gave it (at most) fifteen years. Lederberg also thought that the time was "very near" when "the size of the brain . . . would be brought under direct developmental control," when we could create much bigger, better brains for children.

One of the great problems for a non-scientist in the field is to evaluate the ability of someone like Lederberg to make such predictions outside his real field of expertise. Lederberg sounded perspicacious at the time, and certainly exciting (and Toffler was certainly selling excitement about the future in his book), but

Lederberg ignored countless barriers, such as the ability of the government—if it chose—to ban funding for such research.

And as for Toffler, of course it is the tone that sells a book, especially a tone of impending Armageddon:

> It is important for laymen to understand that Lederberg is by no means a lone worrier in the scientific community. His fears about the biological revolution are shared by many of his scientific colleagues. The ethical, moral, and political questions raised by the new biology simply boggle the mind. Who shall live and who shall die? What is man? Who shall control research into these fields? How shall new findings be applied? Might we not unleash horrors for which man is totally unprepared? In the opinion of many of the world's leading scientists the clock is ticking for a "biological Hiroshima."[14]

Well, not really. And I would like to see the hard data that proved, even then, that "many" of the world's top scientists feared such a future, or that Lederberg's views were not confined to a small, speculative minority. In fact, Lederberg was very alone in going out on a limb with his highly speculative predictions.

In the next paragraph, Toffler quotes E. Hafez (a man who, he tells us, is an "internationally respected biologist") who predicted in 1965 that,

> . . . within a mere ten to fifteen years, a woman will be able to buy a tiny embryo, take it to her doctor, have it implanted in her uterus, carry it for nine months and then give birth to it as though it had been conceived in her own body.

It wasn't until 1978 that Louise Brown was born by in vitro fertilization and the first American IVF baby didn't come until 1980. Only in 1996 did some desperate, infertile couples start to pay young women for eggs that would be fertilized with the husband's sperm for implantation in the older woman. Couples still can't "buy" an embryo.

Toffler next quoted Daniele Petrucci (by the way, all his quotes from Hafez and Petrucci came from a sensationalistic article in *Life* magazine in 1965, so Toffler was taking *Life's* word about the credentials of these men and women, who claimed that artificial wombs are just around the corner):

> Indeed, it will be possible at some point to do away with the female uterus altogether. Babies will be conceived, nurtured and raised to maturity outside the human body. It is clearly only a matter of years before the work begun by Dr. Daniele Petrucci in Bologna . . . makes it

> possible for women to have babies without the discomfort of pregnancy.[15]

Petrucci had claimed to have fertilized a human egg in vitro, grown it for 29 days, and then destroyed it because it was growing as a monster. What Toffler didn't discover then was that the evidence for this claim was never provided by Petrucci and the claim was later dismissed as fraudulent. (This fraud was harmful because it fueled later worries that IVF might produce monstrous babies—a fear also raised about cloning.) And of course, we are nowhere near having a real artificial womb.

In (what we can now see as) a hilarious scenario, Toffler somberly quotes Hafez's suggestion that,

> fertilized eggs might be useful in the colonization of planets. Instead of shipping adults to Mars, we could ship a shoebox full of such cells and grow them into an entire city-size population of humans. Dr. Hafez observes, ". . . why send full-grown men and women aboard space ships? Instead, why not ship tiny embryos, in the care of a competent biologist . . . We miniaturize other spacecraft components. Why not the passengers?"[16]

Of course, Toffler could not resist the standard, dire predictions about eugenics, about a super race, and about state-controlled genetic enhancement. He eagerly quotes a kooky Soviet biologist predicting a "genetic arms race" between the Cold War enemies. For Toffler, "we are hurtling toward the time when we are able to breed both super- and sub-races. . . . We will be able to create super-athletes, girls with super-mammaries. . . ."

All these predictions were presented not as science fiction but as factual predictions. Toffler certainly got a lot of attention, but is his legacy a good one? On the good side, he scared people, and made them realize a lot of change had occurred in a few years. On the other side, he also made people feel that the change was uncontrollable and that we could never go back. In those aspects, his legacy has not been a good one.

Other breathlessly-made predictions haven't come true. In the 1960s, computers were seen as the oppressive agents of the State, but in fact personal computers later created new ways of sharing ideas that helped bring down Communism all over the world. Physician-assisted dying for competent, terminal adults in Holland was predicted to turn that peaceful country into an ethical hell, but the practice has been going on for twenty-five years with hardly any bad results. Abortion has been legal in America for a similar twenty-five years and American society continues to function quite nicely.

All these changes—with computers, assisted reproduction, euthanasia, and abortion—were predicted by various seers to land us on an inexorable slide down the slippery slope. None of them came true. So the lesson here is easy: be wary of slope predictions and don't let them make you fear the changes that may bring you a better future.

Finally, one way that the first and last tests of this chapter are linked is that the slippery slope predictions often assume bad motives in parents. Ostensibly, desires to have children who lack genetic dysfunction and to make one's children as talented, healthy, and lovable as possible, do not seem like the pit at the bottom of a slippery slope—although from the way many pundits talk about the slippery slope, one might think it so.

Conclusion

I have offered four questions to ask when we discuss the ethics of human asexual reproduction. Of course, these tests are applicable to many other issues in ethics. In thinking about originating humans by cloning, we should not think of such origination as being a moral issue unless someone is harmed, not assume that traditional moral rules are always right because the problems they address may change, not assume the worst motives in parents, and not let predictions about slippery slopes make us fear change.

Notes

1. J. B. Schneewind, "Modern Moral Philosophy," in Peter Singer (ed.), *A Companion to Ethics* (Cambridge, Mass.: Blackwell, 1991), 147.
2. Quoted by James Rachels, in his *Can Ethics Provide Answers? and Other Essays in Moral Philosophy* (Lanham, Md.: Rowman & Littlefield, 1997), 8; from Peter Singer, "Sidgwick and Reflective Equilibrium," *Monist 58* (1974): 516.
3. See James Rachels, "Active and Passive Euthanasia," *New England Journal of Medicine* 292 (January 1975), 78–80.
4. Gustav Niebuhr, "Cloned Sheep Stirs Debate on Its Use on Humans," *New York Times,* 1 March 1997.
5. Gustav Niebuhr, "Cloned Sheep. . . ."
6. See Gregory Pence, *Classic Cases in Medical Ethics,* 2nd ed. (New York: McGraw-Hill, 1995), 296.
7. Jay Hughes, Medical College of Wisconsin Medical Ethics listserv discussion, September 4, 1995. Transplanting a lobe of a liver or lung, or one kidney where two are functioning, is not transplanting a renewable resource but it is also not like transplanting a heart, which can only be done if a person is dead while his heart continues to beat.
8. "Overview on Cloning," *Los Angeles Times,* 27 April 1997.
9. *Newsweek,* 7 August, 1978, 71.
10. Gustav Niebuhr, "Cloned Sheep. . . ."
11. Alvin Toffler coined the term in 1965 in an article in Horizon magazine. The quotation is from his later book, *Future Shock* (New York: Bantam Books, 1970), 2.
12. Alasdair McIntyre, *After Virtue* (South Bend, Ind.: Indiana University Press, 1981).
13. Alvin Toffler, *Future Shock,* 198.
14. Alvin Toffler, *Future Shock,* 198.
15. Alvin Toffler, *Future Shock,* 199–200.
16. Alvin Toffler, *Future Shock,* 200.

Part IV

Perspectives from Policy and Law

Perspectives from Policy and Law: Introduction

HUMAN CLONING PRESENTS far-reaching legislative and policy concerns. Sound public policy must strike a reasonable balance between recognizing the sphere of private rights, whether those of individuals to procreate or those of scientists to conduct research, and acting on behalf of the entire community. In this light, would a ban on human cloning research be justified? Or should there be legislation prohibiting human cloning altogether? This judgment can only be made after carefully weighing benefits and harms. Moreover, public policy must be based upon a clear understanding of what cloning is and what it involves.

New reproductive technologies have raised all sorts of legal issues. Will these issues be especially aggravated by human cloning? Or is human cloning less radical than we think? In which case, is prohibition less justifiable? Do we possess a constitutional right to reproductive privacy? If so, to what extent? On the other hand, do we have a right to genetic uniqueness? If so, how does this square with natural identical twins? Moreover, because of a pre-existing genotype, is there a sort of "genetic bondage" that would adversely affect the human clone? Legal as well as moral questions erupt when consent requirements appear to be bypassed, as in the case of using a child as a somatic cell source for cloning. And, given that cloning does not involve a combining of different genes, is the human clone still constitutionally protected? Are there other harms that may warrant either a temporary ban or an all-out prohibition? What about legal rights to control information concerning our own DNA? And again, does the pre-embryo possess any legal status?

Throughout the various positions offered here, there seems to be a common concern, echoed in other perspectives, that is especially relevant in our consumer-driven society. That is, whatever policy is established must ensure that any form of commercialization of human cloning and its research be strictly prohibited. The issues demand reasonable and sound analysis, but it is unclear whether bioethicists have contributed to reasoned discussion or have added fuel to hyped-up fears.

The fifteen-member National Bioethics Advisory Commission (NBAC) was given just ninety days to come up with recommendations regarding human cloning. The commission assessed the most prominent arguments against human cloning, including the potential harms to the child, to the family, and to society as a whole. After also considering supporting arguments such as appeals to individual liberty and scientific freedom to conduct research, the commission reviewed religious positions. The group sensibly recognized the difficulty of formulating public policy that would give fair consideration both to individual freedoms and rights and to social well-being.

Instead of simply supporting the government's current ban on federal funding of human cloning research, the NBAC took a giant step further in recommending legislation that would make attempts at human cloning a federal crime. By the same token, the wording of the commission's report reflected a cautious position in light of the current status of the science, concluding that, at the present time, potential harms seem to outweigh benefits and that a temporary prohibition is therefore in order. At the same time, it strongly urged more discussion of the issues.

In his testimony before a U.S. Senate subcommittee just weeks after the cloning of Dolly, George Annas fully supported a federal ban on human cloning. Not only did he warn the subcommittee of specific dangers in human cloning, but he also proposed that a federal regulatory oversight mechanism, the Human Experimentation Agency, enact and monitor relevant legislation.

For Annas, the single most dangerous feature in human cloning is that it represents replication, not reproduction. It is a radical departure from other types of assisted reproduction in that the clone would have a single genetic parent, and this would undoubtedly affect the meaning of a human being. This concern follows in the wake of reproductive technologies in which establishing the real parent has been problematic, raising a host of legal issues as to parental legal rights and duties. Therefore, contra Pence, Annas holds that support of cloning on the grounds that it is a continuation of the process achieved through *in vitro* fertilization (IVF) is not sound.

Annas argues further that there are no good reasons for human cloning. Employing literary images (*Frankenstein* and *Brave New World*), he describes human cloning as another experiment reflecting the experimenter's hubris at the expense of the subject. Human cloning also leads to the objectification of children, so that they are regarded more as commodities than as independent human beings.

One of Annas's most powerful arguments, and one which we do not often hear, concerns the prospect of viewing a child as a somatic cell source. For Annas, the moral acceptability of human reproduction must incorporate the partners' voluntary consent. Thus cloning a child violates ethical principles since there is no consent by the cloner. In addition, the procedure puts the child to be cloned at risk, given the potential for mishaps and harms.

A good portion of Lori Andrews's article is based upon her commissioned report to the NBAC. Although Andrews asserts that the strongest claim to a constitutional right to reproductive privacy may be made on behalf of those who are infertile, she cautions that such procreative rights do not extend to the right to use *another's* DNA, such as a child's, in order to produce a clone of that child. Furthermore, she points out that the radical uniqueness of human cloning is that there is already a pre-existing genotype, a point not often raised by others. This is different from natural twins who, though identical genotypes, are born as contemporaries. The fact that an identical genotype already exists raises all sorts of challenges regarding uniqueness, perception of uniqueness, and expectations. And since human cloning does not involve the mixing of genes to produce a new genotype, human cloning is not constitutionally protected.

Andrews clearly advocates banning human cloning because it brings about various psychological and physical risks. For instance, the clone is held captive to a sort of "genetic bondage" due to the expectations and comparisons that others, particularly parents, will have of the clone. It also has negative impacts upon other children who are siblings of the clone. In addition, Andrews addresses a novel concern regarding the possible premature aging of the clone, which strikes a nerve now that the sheep Dolly appears to be aging more rapidly than normal. This raises an interesting philosophical-genetic question: What is the true age of the clone? Does the clone inherit the age of the donor cell? Or does she start afresh as do all of us? Can we anticipate a shortened life expectancy with clones? Andrews also considers problems regarding unknown or hidden mutations with cloning. Finally, as to these risks, she raises the analogy of incest. That is, even if human cloning technology could prevent or at least minimize physical dangers, the psychological risks are still persuasive enough to ban human cloning.

Andrews offers another powerful argument in the claims that human cloning violates an individual's right to have some level of control over his or her genetic information. That is, we have privacy rights regarding the disclosure of information about our own DNA. She argues that a clone's authority over his or her own genetic information is not even respected, and that this is a blatant violation of the Fifth Amendment.

In contrast to Andrews's position, John Robertson claims that the projected harms wrought by human cloning are unlikely to occur. As he sees it, human cloning is hardly a radical departure from other reproductive technologies, and, therefore, restrictions on cloning research are unwarranted. Providing a concise account of the cloning process through blastomere separation, he feels that its most likely application is to overcome infertility. At the same time, he contends that other potential uses of cloning such as cloning for purposes of insuring against possible tragedies involving children and cloning with the aim of selecting certain desirable traits are less likely to occur. As to the claim that embryos will necessarily be destroyed as a consequence of blastomere separation, he points out that such is the case in other contexts besides cloning. And since there seems to be a consensus that embryos lack moral status, no rights are violated. By the same token, studies of identical twins indicate that creating contemporary twins poses no real psychological harm. And whatever expectations and perceptions others may have of the delayed twin, the uniqueness of the delayed twin will still be assured since genotype alone does not contribute to human identity. Finally, Robertson insists that there is little reason to suspect that parents will not want to show love and affection for their cloned child.

Given the balance of real benefits, particularly for infertile couples, over imagined harms, Robertson argues that there are no sound reasons for prohibiting embryo splitting research. In his endorsement, Robertson also underscores the constitutional right to reproduce which, in the absence of harm, assumes priority. At the same time, Robertson admits that government intervention may be justified in order to prevent the commercialization and sale of embryos.

With Ruth Macklin's article, we come full circle to ask, Just what do we mean by cloning? Macklin contends that this definition needs to be clarified before drawing up any public policy. Moreover, she tackles head on the question of likely harms to children who come about through cloning. What about the threat to individual identity and uniqueness? For Macklin, the charge itself is weak, particularly if it confuses genetic uniqueness with individual uniqueness. Macklin further claims that the notion of having a right to genetic uniqueness, a case made by Callahan, makes even less sense since, if there is such a right, it is apparently violated in the case of genetic twins.

What about the psychological harm to the child who is cloned? Macklin downplays what she feels are exaggerated scenarios, indicating that it remains difficult to compare so-called harms brought about by cloning with the option of never having been born. As of now, Macklin argues, there is little factual support for the claim that the cloned child would necessarily be harmed, and she concludes that legislation prohibiting human cloning is therefore premature.

Macklin notes that even some bioethicists have contributed to the public's confusion and fears, by offering the same sort of hype found in the media rather than the reasoned response of time and study. Finally, echoing the concern of many others, Macklin stresses the need for strict laws forbidding the commercialization of any aspects of embryo research involving human cloning.

Cloning Human Beings: Report and Recommendations of the National Bioethics Advisory Commission

Rockville, Maryland, June 1997

The National Bioethics Advisory Commission consisted of the following members: Harold T. Shapiro (Chair, Princeton University), Patricia Backlar (Portland State University), Arturo Brito (University of Miami School of Medicine), Alexander M. Capron (University of Southern California Law Center), Eric J. Cassell (Cornell Medical College), R. Alta Charo (University of Wisconsin), James F. Childress (University of Virginia), David R. Cox (Stanford University School of Medicine), Rhetaugh G. Dumas (University of Michigan), Ezekiel J. Emanuel (Harvard Medical School), Laurie M. Flynn (National Alliance for the Mentally Ill), Carol W. Greider (The Johns Hopkins University School of Medicine), Steven H. Holtzman (Millennium Pharmaceuticals Inc.), Bette O. Kramer (Richmond Bioethics Consortium), Bernard Lo (University of California, San Francisco), Lawrence H. Miike (State Department of Health, Honolulu, Hawaii), Thomas H. Murray (The Hastings Center), Diane Scott-Jones (Temple University).

This is the final version of the report Cloning Human Beings, *which includes some editing revisions and changes in page numbering.*

Recommendations of the Commission

WITH THE ANNOUNCEMENT that an apparently quite normal sheep had been born in Scotland as a result of somatic cell nuclear transfer cloning came the realization that, as a society, we must yet again collectively decide whether and how to use what appeared to be a dramatic new technological power. The promise and the peril of this scientific advance was noted immediately around the world, but the prospects of creating human beings through this technique mainly elicited widespread resistance and/or concern. Despite this reaction, the scientific significance of the accomplishment, in terms of improved understanding of cell development and cell differentiation, should not be lost. The challenge to public policy is to support the myriad beneficial applications of this new technology while simultaneously guarding against its more questionable uses.

Much of the negative reaction to the potential application of such cloning in humans can be attributed to fears about harms to the children who may result, particularly psychological harms associated with a possibly diminished sense of individuality and personal autonomy. Others express concern about a degradation in the quality of parenting and family life. And virtually all people agree that the current risks of physical harm to children associated with somatic cell nuclear transplantation cloning justify a prohibition at this time on such experimentation.

In addition to concerns about specific harms to children, people have frequently expressed fears that a widespread practice of somatic cell nuclear transfer cloning would undermine important social values by opening the door to a form of eugenics or by tempting some to manipulate others as if they were objects instead of persons. Arrayed against these concerns are other important social values, such as protecting personal choice, particularly in matters pertaining to procreation and child rearing, maintaining privacy and the freedom of scientific inquiry, and encouraging the possible development of new biomedical breakthroughs.

As somatic cell nuclear transfer cloning could represent a means of human reproduction for some people, limitations on that choice must be made only when the societal benefits of prohibition clearly outweigh the value of maintaining the private nature of such highly personal decisions. Especially in light of some arguably compelling cases for attempting to clone a human being using somatic cell nuclear transfer, the ethics of policy making must strike a balance between the values society wishes to reflect and issues of privacy and the freedom of individual choice.

To arrive at its recommendations concerning the use of somatic cell nuclear transfer techniques, NBAC also examined long-standing religious traditions that often influence and guide citizens' responses to new technologies. Religious positions on human cloning are pluralistic in their premises, modes of argument, and conclusions. Nevertheless, several major themes are prominent in Jewish, Roman Catholic, Protestant, and Islamic positions, including responsible human dominion over nature, human dignity and destiny, procreation, and family life. Some religious thinkers argue that the use of somatic cell nuclear transfer cloning to create a child would be intrinsically immoral and thus could never be morally justified; they usually propose a ban on such human cloning. Other religious thinkers contend that human cloning to create a child could be morally justified under some circumstances but hold that it should be strictly regulated in order to prevent abuses.

The public policies recommended with respect to the creation of a child using somatic cell nuclear transfer reflect the Commission's best judgments about both the ethics of attempting such an experiment and its view of traditions regarding limitations on individual actions in the name of the common good. At present, the use of this technique to create a child would be a premature experiment that exposes the developing child to unacceptable risks. This in itself might be sufficient to justify a prohibition on cloning human beings at this time, even if such efforts were to be characterized as the exercise of a fundamental right to attempt to procreate. More speculative psychological harms to the child, and effects on the moral, religious, and cultural values of society, may be enough to justify continued prohibitions in the future, but more time is needed for discussion and evaluation of these concerns.

Beyond the issue of the safety of the procedure, however, NBAC found that concerns relating to the potential psychological harms to children and effects on the moral, religious, and cultural values of society merited further reflection and deliberation. Whether upon such further deliberation our nation will conclude that the use of cloning techniques to create children should be allowed or permanently banned is, for the moment, an open question. Time is an ally in this regard, allowing for the accrual of further data from animal experimentation, enabling an assessment of the prospective safety and efficacy of the procedure in humans, as well as granting a period of fuller national debate on ethical and social concerns. The Commission therefore concluded that a period of time should be imposed in which no attempt is made to create a child using somatic cell nuclear transfer.

Within this overall framework the Commission came to the following conclusions and recommendations:

I. The Commission concludes that at this time it is morally unacceptable for anyone in the public or private sector, whether in a research or clinical setting, to attempt to create a child using somatic cell nuclear transfer cloning. The Commission reached a consensus on this point because current scientific information indicates that this technique is not safe to use in humans at this time. Indeed, the Commission believes it would violate important ethical obligations were clinicians or researchers to attempt to create a child using these particular technologies, which are likely to involve unacceptable risks to the fetus and/or potential child. Moreover, in addition to safety concerns, many other serious ethical concerns have been identified, which require much more widespread and careful public deliberation before this technology may be used.

The Commission, therefore, recommends the following for immediate action:

- A continuation of the current moratorium on the use of federal funding in support of any attempt to create a child by somatic cell nuclear transfer.
- An immediate request to all firms, clinicians, investigators, and professional societies in the private and non-federally funded sectors to comply voluntarily with the intent of the federal moratorium. Professional and scientific societies should make clear that any attempt to create a child by somatic cell nuclear transfer and implantation into a woman's body would at this time be an irresponsible, unethical, and unprofessional act.

II. The Commission further recommends that:

- Federal legislation should be enacted to prohibit anyone from attempting, whether in a research or clinical setting, to create a child through somatic cell nuclear transfer cloning. It is critical, however, that such legislation include a sunset clause to ensure that Congress will review the issue after a specified time period (three to five years) in order to decide whether the prohibition continues to be needed. If state legislation is enacted, it should also contain such a sunset provision. Any such legislation or associated regulation also ought to require that at some point prior to the expiration of the sunset period, an appropriate oversight body will evaluate and report on the current status of somatic cell nuclear transfer technology and on the ethical and social issues that its potential use to create human beings would raise in light of public understandings at that time.

III. The Commission also concludes that:

- Any regulatory or legislative actions undertaken to effect the foregoing prohibition on creating a child by somatic cell nuclear transfer should be carefully written so as not to interfere with other important areas of scientific research. In particular, no new regulations are required regarding the cloning of human DNA sequences and cell lines, since neither activity raises the scientific and ethical issues that arise from the attempt to create children through somatic cell nuclear transfer, and these fields of research have already provided important scientific and biomedical advances. Likewise, research on cloning animals by somatic cell nuclear transfer does not raise the issues implicated in attempting to use this technique for human cloning, and its continuation should only be subject to existing regulations regarding the humane use of animals and review by institution-based animal protection committees.
- If a legislative ban is not enacted, or if a legislative ban is ever lifted, clinical use of somatic cell nuclear transfer techniques to create a child should be preceded by research trials that are governed by the twin protections of

independent review and informed consent, consistent with existing norms of human subjects protection.

- The United States Government should cooperate with other nations and international organizations to enforce any common aspects of their respective policies on the cloning of human beings.

IV. The Commission also concludes that different ethical and religious perspectives and traditions are divided on many of the important moral issues that surround any attempt to create a child using somatic cell nuclear transfer techniques. Therefore, the Commission recommends that:

- The federal government and all interested and concerned parties encourage widespread and continuing deliberation on these issues in order to further our understanding of the ethical and social implications of this technology and to enable society to produce appropriate long-term policies regarding this technology should the time come when present concerns about safety have been addressed.

V. Finally, because scientific knowledge is essential for all citizens to participate in a full and informed fashion in the governance of our complex society, the Commission recommends that:

- Federal departments and agencies concerned with science cooperate in seeking out and supporting opportunities to provide information and education to the public in the area of genetics and other developments in the biomedical sciences, especially where these affect important cultural practices, values, and beliefs.

Scientific Discoveries and Cloning: Challenges for Public Policy

Testimony before the Subcommittee on Public Health and Safety, Committee on Labor and Human Resources, United States Senate, March 12, 1997

by George J. Annas

George Annas is Professor and Chair of the Department of Health Law at Boston University. He also founded the university's Law, Medicine, and Ethics Program.

SENATOR FRIST, THANK YOU for the opportunity to appear before your subcommittee to address some of the legal and ethical aspects surrounding the prospect of human cloning. I agree with President Clinton that we must "resist the temptation to replicate ourselves" and that the use of federal funds for the cloning of human beings should be prohibited. On the other hand, the contours of any broader ban on human cloning require, I believe, sufficient clarity to permit at least some research on the cellular level. This hearing provides an important opportunity to help explore and define just what makes the prospect of human cloning so disturbing to most Americans, and what steps the federal gov-

---

ernment can take to prevent the duplication of human beings without preventing vital research from proceeding.

I will make three basic points this morning: (1) the negative reaction to the prospect of human cloning by the scientific, industrial and public sectors is correct because the cloning of a human would cross a boundary that represents a difference in kind rather than in degree in human "reproduction"; (2) there are no good or sufficient reasons to clone a human; and (3) the prospect of cloning a human being provides an opportunity to establish a new regulatory framework for novel and extreme human experiments.

1. The Cloning of a Human Would Cross a Natural Boundary That Represents a Difference in Kind Rather Than Degree of Human "Reproduction."

There are those who worry about threats to biodiversity by cloning animals, and even potential harm to the animals themselves. But virtually all of the reaction to the appearance of Dolly on the world stage has focused on the potential use of the new cloning technology to replicate a human being. What is so simultaneously fascinating and horrifying about this technology that produced this response? The answer is simple, if not always well-articulated: *replication of a human by cloning would radically alter the very definition of a human being by producing the world's first human with a single genetic parent.* Cloning a human is also viewed as uniquely disturbing because it is the manufacture of a person made to order, represents the potential loss of individuality, and symbolizes the scientist's unrestrained quest for mastery over nature for the sake of knowledge, power, and profits.

Human cloning has been on the public agenda before, and we should recognize the concerns that have been raised by both scientists and policy makers over the past twenty-five years. In 1972, for example, the House Subcommittee on Science, Research and Development of the Committee on Science and Astronautics asked the Science Policy Research Division of the Library of Congress to do a study on the status of genetic engineering. Among other things, that report dealt specifically with cloning and parthenogenesis as it could be applied to humans. Although the report concluded that the cloning of human beings by nuclear substitution "is not now possible," it concluded that cloning "might be considered an advanced type of genetic engineering" if combined with the introduction of highly desirable DNA to "achieve some ultimate objective of genetic engineering." The Report called for assessment and detailed knowledge, forethought and evaluation of the course of genetic developments, rather than "acceptance of the haphazard evolution of the techniques of genetic engineering [in the hope that] the issues will resolve themselves."

Six years later, in 1978, the Subcommittee on Health and the Environment of the House Committee on Interstate and Foreign Commerce held hearings on human cloning in response to the publication of David Rorvick's *The Cloning of a Man.* All of the scientists who testified assured the committee that the supposed account of the cloning of a human being was fictional, and that the techniques described in the book could not work. One scientist testified that he hoped that by showing that the report was false it would also become apparent that the issue of human cloning itself "is a false one, that the apprehensions people have about cloning of human beings are totally unfounded." The major point the scientists wanted to make, however, was that they didn't want any regulations that might affect their research. In the words of one, "There is no need for any form of regulatory legislation, and it could only in the long run have a harmful effect."

Congressional discussion of human cloning was interrupted by the birth of Baby Louise Brown, the world's first IVF baby, in 1978. The ability to conceive a child in a laboratory not only added a new way (in addition to artificial insemination) for humans to reproduce without sex, but also made it possible for the first time for a woman to gestate and give birth to a child to whom she had no genetic relationship. Since 1978, a child can have at least five parents: a genetic and rearing father, and a genetic, gestational, and rearing mother. We pride ourselves as having adapted to this brave new biological world, but in fact we have yet to develop reasonable and enforceable rules for even so elementary a question as who among these five possible parents the law should recognize as those with rights and obligations to the child. Many other problems, including embryo storage and disposition, posthumous use of gametes, and information available to the child also remain unresolved.

IVF represents a striking technological approach to infertility; nonetheless the child is still conceived by the union of an egg and sperm from two separate human beings of the opposite sex. Even though no change in the genetics and biology of embryo creation and growth is at stake in IVF, society continues to wrestle with fundamental issues involving this method of reproduction twenty years after its introduction. Viewing IVF as a precedent for human cloning misses the point. Over the past two decades many ethicists have been accused of "crying wolf" when new medical and scientific technologies have been introduced. This may have been the case in some instances, but not here. This change in kind in the fundamental way in which humans can "reproduce" represents such a challenge to human dignity and the potential devaluation of human life (even comparing the "original" to the "copy" in terms of which is to be more valued) that even the search for an analogy has come up empty handed.

Cloning is replication, not reproduction, and represents a difference in kind not in degree in the manner in which human beings reproduce. Thus, although the constitutional right not to reproduce would seem to apply with equal force to a right not to replicate, to the extent that there is a constitutional right to reproduce (if one is able to), it seems unlikely that existing privacy or liberty doctrine would extend this right to replication by cloning.

2. There are no good or sufficient reasons to clone a human. When the President's Bioethics Commission reported on genetic engineering in 1982 in their report entitled *Splicing Life,* human cloning rated only a short paragraph in a footnote. The paragraph concluded: "The technology to clone a human does not—and may never—exist. Moreover, the critical nongenetic influences on development make it difficult to imagine producing a human clone who would act or appear 'identical'." (p. 10) The NIH Human Embryo Research panel that reported on human embryo research in September 1994 also devoted only a single footnote to this type of cloning. "Popular notions of cloning derive from science fiction books and films that have more to do with cultural fantasies than actual scientific experiments." (at p. 39) Both of these expert panels were wrong to disregard lessons from our literary heritage on this topic, thereby attempting to sever science from its cultural context.

Literary treatments of cloning help inform us that applying this technology to humans is too dangerous to human life and values. The reporter who described Dr. Ian Wilmut as "Dolly's laboratory father" couldn't have conjured up images of Mary Shelley's *Frankenstein* better if he had tried. Frankenstein was also his creature's father/god; the creature telling him: "I ought to be thy Adam." Like Dolly, the "spark of life" was infused into the creature by an electric current. Unlike Dolly, the creature was created as a fully grown adult (not a cloning possibility, but what many Americans fantasize and fear), and wanted more than creaturehood: he wanted a mate of his "own kind" with whom to live, and reproduce. Frankenstein reluctantly agreed to manufacture such a mate if the creature agrees to leave humankind alone, but in the end, viciously destroyed the female creature-mate, concluding that he has no right to inflict the children of this pair, "a race of devils," upon "everlasting generations." Frankenstein ultimately recognized his responsibilities to humanity, and Shelley's great novel explores virtually all the noncommercial elements of today's cloning debate.

The naming of the world's first cloned mammal also has great significance. The sole survivor of 277 cloned embryos (or "fused couplets"), the clone could have been named after its sequence number in this group (e.g., C-137), but this would have only emphasized its character as a produced product. In stark contrast, the name Dolly (provided for the public and not used in the scientific report in *Nature*) suggests an individual, a human or at least a pet. Even at the manufactured level a "doll" is something that produces great joy in our children and is itself harmless. Victor Frankenstein, of course, never named his creature, thereby repudiating any parental responsibility. The creature himself evolved into a monster when it was rejected not only by Frankenstein, but by society as well. Naming the world's first mammal-clone Dolly is meant to distance her from the Frankenstein myth both by making her appear as something she is not, and by assuming parental obligations toward her.

Unlike Shelley's, Aldous Huxley's *Brave New World* future in which all humans are created by cloning through embryo splitting and conditioned to join one of five worker groups, was always unlikely. There are much more efficient ways of creating killers or terrorists (or even workers) than through cloning—physical and psychological conditioning can turn teenagers into terrorists in a matter of months, rather than waiting some eighteen to twenty years for the clones to grow up and be trained themselves. Cloning has no real military or paramilitary uses. Even Hitler's clone would himself likely be quite a different person because he would grow up in a radically altered world environment.

It has been suggested, however, that there might be good reasons to clone a human. Perhaps most compelling is cloning a dying child if this is what the grieving parents want. But this should not be permitted. Not only does this encourage the parents to produce one child in the image of another, it also encourages all of us to view children as interchangeable commodities, because cloning is so different from human reproduction. When a child is cloned, it is not the parents that are being replicated (or are "reproducing") but the child. No one should have such dominion over a child (even a dead or dying child) as to be permitted to use its genes to create the child's child. Humans have a basic right not to reproduce, and human reproduction (even replication) is not like reproducing farm animals, or even pets. Ethical human reproduction properly requires the voluntary participation of the genetic parents. Such voluntary participation is not possible for a young child. Related human rights and dignity would also prohibit using cloned children as organ sources for their father/mother original. Nor is there any "right to be cloned" that an adult might possess that is triggered by marriage to someone with whom the adult cannot reproduce.

Any attempt to clone a human being should also be prohibited by basic ethical principles that prohibit putting human subjects at significant risk without their informed consent. Dolly's birth was a one in 277 embryo chance. The birth of a human from cloning might be technologically possible, but we could only discover this by unethically subjecting the planned child to the risk of serious genetic or physical injury, and subjecting a planned child to this type of risk could literally never be justified. Because we will likely never be able to protect the human subject of cloning research from serious harm, the basic ethical rules of human experimentation prohibit us from ever using it on humans.

3. Developing a regulatory framework for human cloning

What should we do to prevent Dolly technology from being used to manufacture duplicate humans? We have three basic models for scientific/medical policymaking in the U.S.: the market, professional standards, and legislation. We tend to worship the first, distrust the second, and disdain the third. Nonetheless, the prospect of human cloning requires more deliberation about social and moral issues than either the market or science can provide. The market has no morality, and if we believe important values including issues of human rights and human dignity are at stake, we cannot leave cloning decisions to the market. The Biotechnology Technology Industry Organization in the U.S. has already taken the commendable position that human cloning should be prohibited by law. Science often pretends to be value-free, but in fact follows its own imperatives, and either out of ignorance or self-interest assumes that others are making the policy decisions about whether or how to apply the fruits of their labors. We disdain government involvement in reproductive medicine. But cloning is different in kind, and only government has the authority to restrain science and technology until its social and moral implications are adequately examined.

We have a number of options. The first is for Congress to simply ban the use of human cloning. Cloning for replication can (and should) be confined to nonhuman life. We need not, however, prohibit all possible research at the cellular level. For example, to the extent that scientists can make a compelling case for use of cloning technology on the cellular level for research on processes such as cell differentiation and senescence, and so long as any and all attempts to implant a resulting embryo into a human or other animal, or to continue cell division beyond a 14-day period are prohibited, use of human cells for research could be permitted. Anyone proposing such research, however, should have the burden of proving that the research is vital, cannot be conducted any other way, and is unlikely to produce harm to society.

The prospect of human cloning also provides Congress with the opportunity to go beyond ad hoc bans on procedures and funding, and the periodic appointment of blue ribbon committees, and to establish a Human Experimentation Agency with both rule-making and adjudicatory authority in the area of human experimentation. Such an agency could both promulgate rules governing human research and review and approve or disapprove research proposals in areas such as human cloning [on] which local IRBs are simply incapable of providing meaningful reviews. The President's Bioethics panel is important and useful as a forum for discussion and possible policy development. But we have had such panels before, and it is time to move beyond discussion to meaningful regulation in areas like cloning where there is already a societal consensus.

One of the most important procedural steps a federal Human Experimentation Agency should take is to put the burden of proof on those who propose to do extreme and novel experiments, such as cloning, that cross recognized boundaries and call deeply held societal values into question. Thus, cloning proponents should have to prove that there is a compelling reason to approve research on it. I think the Canadian Royal Commission on New Reproductive Technologies quite properly concluded that both cloning and embryo splitting have "no foreseeable ethically acceptable application to the human situation" and therefore should not be done. We need an effective mechanism to ensure that it is not.

Mom, Dad, Clone: Implications for Reproductive Privacy

by Lori B. Andrews

Lori Andrews is Professor of Law at the University of Chicago–Kent School of Law. She presented testimony to the National Bioethics Advisory Commission (NBAC) concerning legal issues in human cloning.

ON 5 JULY 1996 a sheep named Dolly was born in Scotland, the result of the transfer of the nucleus of an adult mammary tissue cell to the enucleated egg cell of an unrelated sheep, and gestation in a third, surrogate mother sheep.[1] Although for the past ten years scientists have routinely cloned sheep and cows from embryo cells,[2] this was the first cloning experiment that apparently succeeded using the nucleus of an adult cell.[3]

Shortly after the report of the sheep cloning was published, President Clinton instituted a ban on federal funding for human cloning,[4] and asked the National Bioethics Advisory Commission (NBAC) to analyze the scientific, legal, and ethical status of human cloning and to make policy recommendations. In June 1997 NBAC recommended the passage of a federal statute that would, for a period of three to five years, ban the implantation of embryos created through human cloning, whether using private or public funding. President Clinton forwarded a bill to Congress prohibiting creating children through human cloning in the United States for at least five years.

If such a law were passed, it might be challenged as violating an individual's or a couple's right to create a biologically related child. This article explores whether such a right exists and whether, even if it does, a ban on creating children through cloning should nonetheless be upheld.

The Right to Make Reproductive Decisions

The right to make decisions about whether to bear children is constitutionally protected under the constitutional right to privacy[5] and the constitutional right to liberty.[6] The U.S. Supreme Court in 1992 reaffirmed the "recognized protection accorded to liberty relating to intimate relationships, the family, and decisions about whether to bear and beget a child."[7] Early decisions protected married couples' right to privacy to make procreative decisions, but later decisions focused on individuals' rights as well. The U.S. Supreme Court, in *Eisenstadt v. Baird,* stated, "[i]f the right of privacy means anything, it is the right of the *individual,* married or single, to be free from unwarranted governmental intrusion into matters so fundamentally affecting a person as the decision whether to bear or beget a child."[9]

A federal district court has indicated that the right to make procreative decisions encompasses the right of an infertile couple to undergo medically assisted reproduction, including in vitro fertilization and the use of a donated embryo.[10] *Lifchez v. Hartigan*[11] held that a ban on research on conceptuses was unconstitutional because it impermissibly infringed upon a woman's fundamental right to privacy. Although the Illinois statute banning embryo and fetal research at issue in the case permitted in vitro fertilization, it did not allow embryo donation, embryo freezing, or experimental prenatal diagnostic procedures. The court stated:

> It takes no great leap of logic to see that within the cluster of constitutionally protected choices that includes the right to have access to contraceptives, there must be included within that cluster the right to submit to a medical procedure that may bring about, rather than prevent, pregnancy. Chorionic villi sampling is similarly protected. The cluster of constitutional choices that includes the right to abort a fetus within the first trimester must also include the right to submit to a procedure designed to give information about that fetus which can then lead to a decision to abort.[12]

Procreative freedom has been found to protect individuals' and couples' decisions to use contraception, abortion, and existing reproductive technology. Some commentators argue that the U.S. Constitution similarly protects the right to create a child through cloning.

There are a variety of scenarios in which such a right might be asserted. If both members of a couple are infertile, they may wish to clone one or the other of themselves.[13] If one member of the couple has a genetic disorder that the couple does not wish to pass on to a child, they could clone the unaffected member of the couple. In addition, if both husband and wife are carriers of a debilitating recessive genetic disease and are unwilling to run the 25% risk of bearing a child with the disorder, they may seek to clone one or the other of them.[14] This may be the only way in which the couple will be willing to have a child that will carry on their genetic line.

Even people who could reproduce coitally may desire to clone for a variety of reasons. People may want to clone themselves, deceased or living loved ones, or individuals with favored traits. A wealthy childless individual may wish to clone himself or herself to have an heir or to continue to control a family business. Parents who are unable to have another child may want to clone their dying child.[15] This is similar to the current situation in which a couple whose daughter died is making arrangements to have a cryopreserved in vitro embryo created with her egg and donor sperm implanted in a surrogate mother in an attempt to recreate their daughter.[16]

Additionally, an individual or couple might choose to clone a person with favored traits. Respected world figures and celebrities such as Mother Teresa, Michael Jordan, and Michelle Pfeiffer have been suggested as candidates for cloning. Less well-known individuals could also be cloned for specific traits. For example, people with a high pain threshold or resistance to radiation could be cloned.[17] People who can perform a particular job well, such as soldiers, might be cloned.[18] One biologist suggested cloning legless men for the low gravitational field and cramped quarters of a spaceship.[19]

Cloning also offers gay individuals a chance to procreate without using nuclear DNA from a member of the opposite sex. Clone Rights United Front, a group of gay activists based in New York, have been demonstrating against a proposed New York law that would ban nuclear transplantation research and human cloning. They oppose such a ban because they see human cloning as a significant means of legitimizing "same-sex reproduction."[20] Randolfe Wicker founded the Clone Rights United Front in order to pressure legislators not to ban human cloning research because he sees nuclear transplantation cloning as an inalienable reproductive right.[21] Wicker stated, "We're fighting for research, and we're defending people's reproductive rights. . . . I realize my clone would be my identical twin, and my identical twin has a right to be born."[22]

Ann Northrop, a columnist for the New York gay newspaper *LGNY,* says that nuclear transplantation is enticing to lesbians because it offers them a means of reproduction and has the potential of giving women complete control over reproduction.[23] "This is sort of the final nail in men's coffins," she says. "Men are going to have a very hard time justifying their existence on this planet, I think. Maybe women may not let men reproduce."[24]

The strongest claim for procreative freedom is that made by infertile individuals, for whom this is the only way to have a child with a genetic link to them. However, the number of people who will actually need cloning is quite limited. Many people can be helped by in vitro fertilization and its adjuncts; others are comfortable using a donated gamete. In all the other instances of creating a child through cloning, the individual is biologically able to have a child of his or her own, but is choosing not to because he or she prefers to have a child with certain traits. This made-to-order child-making is less compelling than the infertility scenario. Moreover, there is little legal basis to suggest that a person's procreative freedom includes a right to procreate using *someone else's* DNA, such as relatives, or a celebrity. Courts are particularly unlikely to find that parents have a right to clone their young child. Procreative freedom is not a predatory right that would provide access to another individual's DNA.

The right of procreation is likely to be limited to situations in which an individual is creating a biologically related child. It could be argued that cloning oneself invokes that right to an even greater degree than normal reproduction. As lawyer Francis Pizzulli points out, "[i]n comparison with the parent who contributes half of the sexually reproduced child's genetic formula, the clonist is conferred with more than the requisite degree of biological parenthood, since he is the sole genetic parent."[25]

John Robertson argues that cloning is not qualitatively different from the practice of medically assisted reproduction and genetic selection that is currently occurring.[26] Consequently, he argues that "cloning . . . would appear to fall within the fundamental freedom of married couples, including infertile married couples to have biologically related offspring."[27] Similarly, June Coleman argues that the right to make reproductive decisions includes the right to decide in what manner to reproduce, including reproduction through, or made possible by, embryo cryopreservation and embryo twinning.[28] This argument could also be applied to nuclear transplantation by saying that a ban on cloning as a method of reproduction is tantamount to the state denying one's right to reproductive freedom.

In contrast, George Annas argues that cloning does not fall within the constitutional protection of reproductive decisions. "Cloning is replication, not reproduction, and represents a difference in kind, not in degree in the way humans continue the species."[29] He explains that "[t]his change in kind in the fundamental way in which humans can 'reproduce' represents such a challenge to human dignity and the potential devaluation of human life (even comparing the 'original' to the 'copy' in terms of which is to be more valued) that even the search for an analogy has come up empty handed."[30]

The process and resulting relationship created by cloning is profoundly different from that created through normal reproduction or even from that created through reproductive technologies such as in vitro fertilization, artificial insemination, or surrogate motherhood. In even the most high-tech reproductive technologies available, a mix of genes occurs to create an individual with a genotype that has never before existed on earth. In the case of twins, two such individuals are created. Their futures are open and the distinction between themselves and their parents is acknowledged. In the case of cloning, however, the genotype has already existed. Even though it is clear that the individual will develop into a person with different traits because of different social, environmental, and generational influences, there is evidence that the fact that he or she has a genotype that already existed will affect how the resulting clone is treated by himself, his family, and social institutions.

In that sense, cloning is sufficiently distinct from traditional reproduction or alternative reproduction to not be considered constitutionally protected. It is not a process of genetic mix, but of genetic duplication. It is not reproduction, but a sort of recycling, where a single individual's genome is made into someone else.

Assuming Constitutional Protection

Let us assume, though, that courts were willing to make a large leap and find that the constitutional privacy and liberty protections of reproduction encompass cloning. If a constitutional right to clone was recognized, any legislation that would infringe unduly upon this fundamental right would be subject to a "strict standard" of judicial review.[31] Legislation prohibiting the ability to clone or prohibiting research would have to further a compelling interest in the least restrictive manner possible in order to survive this standard of review.[32]

The potential physical and psychological risks of cloning an entire individual are sufficiently compelling to justify banning the procedure. There are many physical risks to the resulting child. Of 277 attempts, only one sheep lived. The high rate of laboratory deaths may suggest that cloning in fact damages the DNA of a cell. In addition, scientists urge that Dolly should be closely monitored for abnormal genetic anomalies that did not kill her as a fetus but may have long-term harmful effects. [33]

For example, all of the initial frog cloning experiments succeeded only to the point of the amphibian's tadpole stage.[34] In addition, some of the tadpoles were grossly malformed. Initial trials in human nuclear transplantation could also meet with disastrous results. Ian Wilmut and National Institutes of Health director Harold Varmus, testifying before Congress, specifically raised the concern that cloning technology is not scientifically ready to be applied to humans, even if it were permitted, because there are technical questions that can only be answered by continued animal research.[35] Dr. Wilmut is specifically concerned with the ethical issue that would be raised by any "defective births," which may be likely to occur if nuclear transplantation is attempted with humans.[36]

In addition, if all the genes in the adult DNA are not properly reactivated, there might be a problem at a later developmental stage in the resulting clone.[37] Some differentiated cells rearrange a subset of their genes. For example, immune cells rearrange some of their genes to make surface molecules.[38] That rearrangement could cause physical problems for the resulting clone.

Also, because scientists do not fully understand the cellular aging process, scientists do not know what 'age' or 'genetic clock' Dolly inherited.[39] On a cellular level, when the *Nature* article was published about her, was she a normal seven-month-old lamb, or was she six years old (the age of the mammary donor cell)?[40] Colin Stewart believes that Dolly's cells most likely are set to the genetic clock of the nucleus donor, and therefore are comparable to those of her six-year-old progenitor.[41] One commentator stated that if the hypotheses of a cellular, self-regulating genetic clock are correct, clones would be cellularly programmed to have much shorter life spans than the "original," which would seriously undermine many of the benefits that have been set forth in support of cloning—mostly agricultural justifications—and would psychologically lead people to view cloned animals and humans as short-lived, disposable copies.[42] This concern for premature aging has lead Dr. Sherman Elias, a geneticist and obstetrician at the Baylor College of Medicine, to call for further animal testing of nuclear transplantation as a safeguard to avoid subjecting human clones to premature aging and the potential harms associated with aged cells.[43]

The hidden mutations that may be passed on by using an adult cell raise concerns as well. Mutations are "a problem with every cell, and you don't even know where to check for them," notes Ralph Brinster of the University of Pennsylvania.[44] "If a brain cell is infected with a mutant skin gene, you would not know because it would not affect the way the cell develops because it is inactive. If you choose the wrong cell, then mutations would become apparent."[45]

When Physical Risks Decline

The proposed federal bill would put a five-year moratorium on creating a child through cloning. During that time period, though, the physical risks of cloning will probably diminish. Animal researchers around the world are rushing to try the Wilmut technique in a range of species. If cloning appeared to be physically safe and reached a certain level of efficiency, should it then be permissible in humans?

The NBAC recommendations left open the possibility of continuing the ban on human cloning based on psychological and social risks.[46] The notion of replicating existing humans seems to fundamentally conflict with our legal system, which emphatically protects individuality and uniqueness.[47] Banning procreation through nuclear transplantation is justifiable in light of the sanctity of the individual and personal privacy notions that are found in different constitutional amendments, and protected by the Fourteenth Amendment.[48]

The clone has lost the ability to control disclosure of intimate personal information. A ban on cloning would "preserve the uniqueness of man's personality and thus safeguard the islands of privacy which surround individuality."[49]

These privacy rights are implicated through a clone's right to "retain and control the disclosure of personal information—foreknowledge of the clonant's genetic predispositions." [50] Catherine Valerio Barrad argues that courts should recognize a privacy interest in one's DNA because science is increasingly able to decipher and gather personal information from one's genetic code.[51] The fear that potential employers and health insurers may use one's private genetic information discriminatorily is not only a problem for the original DNA possessor, but any clone "made" from that individual.[52] Even in cases in which the donor waives his privacy rights and releases genetic information about himself, the privacy rights of the clone are necessarily implicated due to the fact that the clone possesses the same nucleic genetic code.[53] This runs afoul of principles behind the Fifth Amendment's protection of a "person's ability to regulate the disclosure of information about himself."[54]

If a cloned person's genetic progenitor is a famous musician or athlete, parents may exert an improper amount of coercion to get the child to develop those talents. True, the same thing may happen—to a lesser degree—now, but the cloning scenario is more problematic. A parent might force a naturally conceived child to practice piano hours on end, but will probably eventually give up if the child seems disinterested or tone deaf. More fervent attempts to develop the child's musical ability will occur if the parents chose (or even paid for) nuclear material from a talented pianist. And pity the poor child who is the clone of a famous basketball player. If he breaks his kneecap at age ten, will his parents consider him worthless? Will he consider himself a failure?

In attempting to cull out from the resulting child the favored traits of the loved one or celebrity who has been cloned, the social parents will probably limit the environmental stimuli that the child is exposed to. The pianist's clone might not be allowed to play baseball or just hang out with other children. The clone of a dead child might not be exposed to food or experiences that the first child had rejected. The resulting clone may be viewed as being in a type of "genetic bondage"[55] with improper constraints on his or her freedom.

Some scientists argue that this possibility will not come to pass because everyone knows that a clone will be different from the original. The NBAC report puts it this way: "Thus the idea that one could make through somatic cell nuclear transfer a team of Michael Jordans, a physics department of Albert Einsteins, or an opera chorus of Pavarottis, is simply false." [56] But this overlooks the fact that we are in an era of genetic determinism, in which newspapers daily report the gene for this or that and top scientists tell us that we are a packet of genes unfolding.

James Watson, co-discoverer of deoxyribonucleic acid (DNA) and the first director of the Human Genome Project, has stated, "We used to think our fate was in the stars. Now we know, in large measure, our fate is in our genes."[57] Harvard zoologist Edward O. Wilson asserts that the human brain is not *tabula rasa* later filled in by experience but, "an exposed negative waiting to be slipped into developer fluid."[58] Genetics is alleged to be so important by some scientists that it caused psychiatrist David Reiss at George Washington University to declare that "the Cold War is over in the nature and nurture debate."[59]

Whether or not this is true, parents may raise the resulting clone as if it were true. After all, the only reason people want to clone is to assure that the child has a certain genetic makeup. Thus it seems absurd to think they will forget about that genetic makeup once the child comes into being. Elsewhere in our current social policies, though, we limit parents' genetic foreknowledge of their children because we believe it will improperly influence their rearing practices.

Cloning could undermine human dignity by threatening the replicant's sense of self and sense of autonomy. A vast body of developmental psychology research has signaled the need of children to have a sense of an independent self. This might be less likely to occur if they were the clones of a member of the couple raising them or of previous children who died.

The replicant individual may be made to feel that he is less of a free agent. Laurence Tribe argues that if one's genetic makeup is subject to prior determination, "one's ability to conceive of oneself as a free and rational being entitled to resist various social claims may gradually weaken and might finally disappear altogether."[60] Under such an analysis, it does not matter whether genetics actually determines a person's characteristics. Having a predetermined genetic makeup can be limiting if the person rearing the replicant and/or the replicant believes in genetic determinism.[61] In addition, there is much research on the impact of genetic information that demonstrates that a person's genetic foreknowledge about himself or herself (whether negative or positive) can threaten that individual's self-image, harm his or her relationships with family members, and cause social institutions to discriminate against him or her.[62]

Even though parents have a constitutional right to make childrearing decisions similar to their constitutional right to make childbearing decisions, parents do not have a right to receive genetic information about their children that is not of immediate medical benefit. The main concern is that a child about whom genetic information is known in advance will be limited in his or her horizons. A few years ago, a mother entered a Huntington disease testing facility with her two young children. "I'd like you to test my children for the HD gene," she said. "Because I only have enough money to send one to Harvard."[63] That request and similar requests to test young girls for the breast cancer gene or other young children for carrier status for recessive genetic disorders raise concerns about whether parents' genetic knowledge about their child will cause them to treat that child differently. A variety of studies have suggested that there may be risks to giving parents such information.

" 'Planning for the future,' perhaps the most frequently given reason for testing, may become 'restricting the future' (and also the present) by shifting family resources away from a child with a positive diagnosis," wrote Dorothy Wertz, Joanna Fanos, and Philip Reilly, in an article in the *Journal of the American Medical Association.*[64] Such a child "can grow up in a world of limited horizons and may be psychologically harmed even if treatment is subsequently found for the disorder."[65] A joint statement by the American Society of Human Genetics (ASHG) and the American College of Medical Genetics (ACMG) notes, "Presymptomatic diagnosis may preclude insurance coverage or may thwart long term goals such as advanced education or home ownership."[66]

The possibility that genetic testing of children can lead to a dangerous self-fulfilling prophecy led to the demise of one study involving testing children. Harvard researchers proposed to test children to see if they had the XYY chromosomal complement, which had been linked (by flimsy evidence) to criminality. They proposed to study the children for decades to see if those with that genetic makeup were more likely to engage in a crime than those without it. They intended to tell the mothers which children had XYY. Imagine the effect of that information—on the mother, and on the child. Each time the child took his little brother's toy, or lashed out in anger at a playmate, the mother might freeze in horror at the idea that her child's genetic predisposition was unfolding itself. She might intervene when other mothers would normally not, and thus distort the rearing of her child.

Because of the potential psychological and financial harm that genetic testing of children may cause, a growing number of commentators and advisory bodies have recommended that parents not be able to learn genetic information about their children. The Institute of Medicine Committee on Assessing Genetic Risks recommended that "in the clinical setting, children generally be tested only for disorders for which a curative or preventive treatment exists and should be instituted at that early stage. Childhood screening is not appropriate for carrier status, untreatable childhood diseases, and late-onset diseases that cannot be prevented or forestalled by early treatment."[67] The American Society of Human Genetics and American College of Medical Genetics made similar recommendations.

A cloned child will be a child who is likely to be exposed to limited experiences and limited opportunities. Even if he or she is cloned from a person who has favored traits, he may not get the benefit of that heritage. His environment might not provide him with the drive that made the original succeed. Or so many clones may be created from the favored original that their value and opportunities may be lessened. (If the entire NBA consisted of Michael Jordan clones, the game would be far less interesting and each individual less valuable.) In addition, even individuals with favored traits may have genes associated with diseases that could lead to insurance discrimination against the individuals cloned. If Jordan died young of an inheritable cardiac disorder, his clones would find their futures restricted. Banning cloning would be in keeping with philosopher Joel Feinberg's analysis that children have a right to an "open future."[68]

Some commentators argue that potential psychological and social harms from cloning are too speculative to provide the foundation for a government ban. Elsewhere, I have argued that speculative harms do not provide a sufficient reason to ban reproductive arrangements such as in vitro fertilization or surrogate motherhood.[69] But the risks of cloning go far beyond the potential psychological risks to the original whose expectations are not met in the cloning, or the risks to the child of having an unusual family arrangement if the original was not one of his or her rearing parents.

The risk here is of hubris, of abuse of power. Cloning represents the potential for "[a]buses of the power to control another person's destiny—both psychological and physical—of an unprecedented order."[70] Francis Pizzulli points out that legal discussions of whether the replicant is the property of the cloned individual, the same person as the cloned individual, or a resource for organs all show how easily the replicant's own autonomy can be swept aside.[71]

In that sense, maybe the best analogy to cloning is incest. Arguably, reproductive privacy and liberty are threatened as much by a ban on incest as by a ban on cloning. Arguably the harms are equally speculative. Yes, incest creates certain potential physical risks to the offspring, due to the potential for lethal recessive disorders to occur. But no one seriously thinks that this physical risk is the reason we ban incest. Arguably a father and daughter could avoid that risk by contracepting or agreeing to have prenatal diagnosis and abort affected fetuses. There might even be instances in which, because of their personalities, there is no psychological harm to either party.

Despite the fact that risks are speculative—and could be counterbalanced in many cases by other measures—we can ban incest because it is about improper parental power over children. We should ban the cloning of human beings through somatic cell nuclear transfer—even if physical safety is established—for that same reason.

Notes

1. Specter M, Kolata G. A new creation: the path to cloning—a special report. *New York Times* 1997; Mar 3:A1.
2. In 1993, embryologists at George Washington University split human embryos, making twins and triplets. See Sawyer K. Researchers clone human embryo cells; work is small step in aiding infertile. *Washington Post* 1993; Oct 25:A4. These embryos were not implanted into a woman for gestation. This procedure is distinguishable from cloning by nuclear transfer.
3. Begley S. Little lamb, who made thee? *Newsweek* 1997; Mar 10:53–7. See also Wilmut I, Schnieke AE, McWhir J, Kind AJ, Campbell KHS. Viable offspring derived from fetal and adult mammalian cells. *Nature* 1997;385:810–3.
4. Transcript of Clinton remarks on cloning. *U.S. Newswire* 1997;Mar 4.
5. E.g., Griswold v. Connecticut, 381 U.S. 379 (1965); Eisenstadt v. Baird, 405 U.S. 438 (1972).
6. Planned Parenthood v. Casey, 505 U.S. 833,112 S.Ct. 2791 (1992).

7. Planned Parenthood v. Casey, 505 U.S. 833,112 S.Ct. 2791, 2810 (1992).
8. Eisenstadt v. Baird, 405 U.S. 438 (1972).
9. Eisenstadt v. Baird, 405 U.S. 438,453 (1972).
10. Lifchez v. Hartigan, 735 F.Supp. 1361 (N.D. Ill.), aff'd without opinion, *sub nom.;* Scholberg v. Lifchez, 914 F.2d 260 (7th Cir. 1990). cert. denied, 111 S.Ct. 787 (1991).
11. See note 10, Lifchez v. Hartigan 1991.
12. See note 10, Lifchez v. Hartigan 1991:1377 (citations omitted). The court also held that the statute was impermissibly vague because of its failure to define "experiment" or "therapeutic" (at 1376).
13. See Wray H, Sheler JL, Watson T. The world after cloning. *U.S. News & World Report* 1997;Mar 10:59.
14. Katz J. *Experimentation with Human Beings* 977 (1972).
15. Gaylin W. We have the awful knowledge to make exact copies of human beings. *New York Times* 1997;Mar 5:48.
16. Kolata G. Medicine's troubling bonus: surplus of human embryos, *New York Times* 1997;Mar 16:1. "Fox on Trends," Fox Television Broadcast, 19 March 1997.
17. Haldane JBS. Biological possibilities for the human species in the next thousand years. In Wolstenholme G, ed. *Man and His Future.* London: Churchill, 1963:337. Cited in Pizzulli FC. Asexual reproduction and genetic engineering: a constitutional assessment of the technology of cloning [Note]. *Southern California Law Review* 1974;47:490, n. 66.
18. Fletcher J. Ethical aspects of genetic controls. *New England Journal of Medicine* 1971;285:779.
19. See note 17. Pizzulli 1974:520.
20. Manning A. Pressing a 'right' to clone humans, some gays foresee reproduction option. *USA Today* 1997;Mar 6:D1.
21. See note 20. Manning 1997; see also Schilinger L. Postcard from New York. *The Independent* [London] 1997;Mar 16:2.
22. See note 21, Schilinger 1997.
23. See note 20. Manning 1997.
24. See note 21. Schilinger 1997.
25. See note 17. Pizzulli 1974:550. Charles Strom, director of genetics and the DNA laboratory at Illinois Masonic Medical Center argues that the high rate of embryo death that has occurred in animal cloning should not dissuade people from considering cloning as a legitimate reproductive technique. Strom points out that all new reproductive technologies have been marred by high failure rates and that it is just a matter of time before cloning could be as economically efficient as any other form of artificial reproduction. See Stolberg S. Sheep clone researcher calls for caution science. *Los Angeles Times* 1997;Mar 1:A18.
26. Robertson J. Statement to the National Bioethics Advisory Commission. 14 March 1997:83. This seems to be a reversal of Robertson's earlier position that cloning "may deviate too far from prevailing conception of what is valuable about reproduction to count as a protected reproductive experience. At some point attempts to control the entire genome of a new person pass beyond the central experiences of identity and meaning that make reproduction a valued experience." Robertson J. *Children of Choice: Freedom and the New Reproductive Technologies.* Princeton, New Jersey: Princeton University Press, 1994:169.
27. See note 26, Robertson 1994.
28. See Coleman J. Playing God or playing scientist: a constitutional analysis of laws banning embryological procedures. *Pacific Law Journal* 1996;27:1351.
29. Annas GJ. Human cloning. *ABA Journal* 1997;13:80–81.
30. Annas GJ. Testimony on scientific discoveries and cloning: challenges for public policy. Subcommittee on Public Health and Safety, Committee on Labor and Human Resources, United States Senate. 12 March 1977:4.
31. See. e.g., Griswold v. Connecticut, 381 U.S. 479 (1965); Eisenstadt v. Baird, 405 U.S. 438 (1972); Roe v. Wade, 410 U.S. 113 (1973); Planned Parenthood of Southern Pennsylvania v. Casey, 505 U.S. 833 (1992).

32. See note 10, Lifchez v. Hartigan.
33. See Nash JM. The age of cloning. *Time* 1997; Mar 10:62–65; see also Spotts PN, Marquand R. A lamb ignites a debate on the ethics of cloning. *Christian Science Monitor* 1997; Feb. 26:3.
34. See The law and medicine. *The Economist* 1997;Mar 1:59; see also note 17, Pizzulli 1974:484.
35. See Recer P. Sheep cloner says cloning people would be inhumane. *Associated Press* 1997;Mar. 12. Reported testimony of Dr. Ian Wilmut and of Dr. Harold Varmus before the Senate on 12 March 1997 regarding the banning of human cloning research.
36. See note 35, Recer 1997. Comments of Dr. Ian Wilmut, testifying that as of yet he does not know of "any reason why we would want to copy a person. I personally have still not heard of a potential use of this technique to produce a new person that I would find ethical or acceptable."
37. Tilghman S. Statement to National Bioethics Advisory Commission, 13 March 1997:173.
38. See note 37, Tilghman 1997:147.
39. See note 35, Recer 1997.
40. See note 35, Recer 1997; see also note 33, Nash 1997:62–65.
41. See note 35, Recer 1997; Laurence J, Hornsby M. Warning on human clones. *Times* [London]1997;Feb 23. Whatever next? *The Economist* 1997; Mar 1:79 (discussing the problems associated with having mitochondria of egg interact with donor cell).
42. Hello Dolly. *The Economist* 1997;Mar 1:17, discussing the pros and cons of aging research that could result from nuclear transplantation cloning; cf. Monmaney T. Prospect of human cloning gives birth to volatile issues. *Los Angeles Times* 1997;Mar 2:A2.
43. See note 42, Monmaney 1997.
44. See note 33, Nash 1997.
45. See note 33, Nash 1997; see also note 37, Tilghman 1997:145.
46. National Bioethics Advisory Commission. *Cloning Human Beings: Report and Recommendations of the National Bioethics Advisory Commission.* Rockville, Maryland: National Bioethics Advisory Commission, 1997:9.
47. Mauro T. Sheep clone prompts U.S. panel review. *USA Today* 1997; Feb 25:A1.
48. See note 17, Pizzulli 1974:502.
49. See note 17, Pizzulli 1974:512.
50. See note 17, Pizzulli 1974. See also Amer MS, Breaking the mold: human embryo cloning and its implications for a right to individuality. *UCLA Law Review* 1996;4:1666.
51. Valerio Barrad CM. Genetic information and property theory [Comment]. *Northwestern University Law Review* 1993;87:1050.
52. See note 51, Valerlo Barrad 1993.
53. See note 51, Valerlo Barrad 1993.
54. See note 50. Amer 1996.
55. See note 17, Pizzulli 1974.
56. See note 46, National Bioethics Advisory Commission 1997:33.
57. Jaroff L. The gene hunt. *Time* 1989;Mar 20:63.
58. Wolfe T. Sorry, But Your Soul Just Died. *Forbes ASAP* 1996; Dec 2:212.
59. Mann CC. Behavioral genetics in transition. *Science* 1994;264:1686.
60. Tribe L. Technology assessment and the fourth discontinuity: the limits of instrumental rationality. *Southern California Law Review* 1973;46:648.
61. There is much evidence of the widespread belief in genetic determinism. See. e.g.. Nelkin D, Lindee MS, *The DNA Mystique: The Gene as Cultural Icon.* New York: W.H. Freeman & Company, 1995.
62. For a review of the studies, see Andrews LB. Prenatal screening and the culture of motherhood. *Hastings Law Journal* 1996;47:967.
63. Wexler N. Clairvoyance and caution: repercussions from the Human Genome Project. In Keyles DJ, Hood L. *The Code of Codes: Scientific and Social Issues in the Human Genome Project.* Cambridge, Massachusetts: Harvard University Press, 1992:211–43, 233.
64. Wertz D, Fanos J., Reilly P. Genetic testing for children and adolescents: who decides? *JAMA* 1994;274:878.

65. See note 64, Wertz et al. 1994. Similarly, the ASHG/ACMG Statement notes: "Expectations of others for education, social relationships and/or employment may be significantly altered when a child is found to carry a gene associated with a late-onset disease or susceptibility. Such individuals may not be encouraged to reach their full potential, or they may have difficulty obtaining education or employment if their risk for early death or disability is revealed." American Society of Human Genetics and American College of Medical Genetics. Points to consider: ethical, legal, and psychosocial implications of genetic testing in children and adolescents. *American Journal of Human Genetics* 1995;57:1233–41, 1236.
66. See note 65, ASHG/ACMG, 1995.
67. Andrews, LB, Fullerton JE, Holtzman NA, Motulsky AG, eds. *Assessing Genetic Risks: Implications for Health and Social Policy.* Washington, D.C.: National Academy of Sciences, 1994:276.
68. See note 46, National Bioethics Advisory Commission 1997:67.
69. Andrews LB. Surrogate motherhood: the challenge for feminists. *Law, Medicine & Health Care* 1988;72:16.
70. See note 17, Pizzolli 1974:497.
71. See note 17, Pizzolli 1974:492.

The Question of Human Cloning

by John A. Robertson

John A. Robertson is the Thomas Watt Gregory Professor of Law, University of Texas at Austin.

THE IDEA OF splitting off cells from embryos to clone human beings sounds so bizarre and dangerous that one would think the practice should not be permitted. A closer look reveals its ethical acceptability.

Accustomed though we are to advances in medical technology, a 24 October 1993 news report that human embryos had been cloned astonished many persons. A *New York Times* story, "Researcher Clones Embryos of Humans in Fertility Effort," was the feature that Sunday morning in many newspapers throughout the country. Media coverage continued for several days, with debates about cloning on editorial pages, *Nightline* and *Larry King Live*.

Within a week the issue had faded from media consciousness, aided in part by *Time* and *Newsweek* stories that stressed the huge gap between the reported research and the *Jurassic Park*-type fears of cloned human beings that initially spurred national coverage.[1] Bioethicists and lawmakers, however, must still contend with the ethical and policy issues that even limited cloning of humans presents. Should researchers be free to continue cloning research? May infertile couples and their physicians employ cloning to form families? Or should government prevent cloning research or discourage some or all of its later applications?

John A. Robertson, "The Question of Human Cloning," *Hastings Center Report*, vol. 24, no. 2 (March–April 1994): 6–14. Reprinted by permission of the Hastings Center and the author.

As with many biomedical developments, these questions present a mix of issues that need careful sorting. They involve, among others, questions of the propriety of embryo research, the validity of deliberately creating twins, and the importance of nature versus nurture in forming human beings. They also raise slippery slope concerns: should otherwise seemingly valid uses of a new technique be stopped to prevent later undesirable uses from occurring? To address those issues we must first describe the cloning research that has touched off the furor and the concerns that it presents.

Two Types of Cloning

The research that put cloning on the public agenda was a long way from Huxleyian fantasies of identical babies, mass produced in laboratories, and did not involve cloning as conventionally understood at all. To clone means to create a genetic copy or replica. Perhaps due to science fiction fantasies, it has been assumed that cloning would occur by removing the nucleus from the cell of one person, placing it in an egg that has had its nucleus removed, and then implanting it in a laboratory incubator or a woman who would bring to term a child with the identical genetic characteristics of the person providing the cell nucleus. Although this procedure has worked with frogs, it has never succeeded with mammals and appears highly unlikely to be accomplished in even the midrange future. If this form of cloning were possible, scientists could fabricate as many copies as one wished of any available human genome, subject only to the limits of uterine or artificial gestation.

A second and more limited way to create clones is to split the cells or blastomeres of an early multicelled embryo before the cells have begun to differentiate. Because each blastomere at this stage is in theory totipotent (that is, capable of producing an entire organism itself), the separated cells can become new embryos, all of which will have the same genome. This form of cloning is now practiced to some extent in the cattle industry. Cloning by blastomere separation is limited to the number of cells that can be separated before cell differentiation, which destroys totipotency, occurs.

The study that generated the recent interest in cloning involved a small but essential step toward cloning human beings by embryo splitting. Researchers at George Washington University Hospital in Washington, D.C., separated cells or blastomeres from seventeen two- to eight-celled preembryos and showed that, to a limited extent, they would divide and grow in culture. The cells had been obtained from polyspermic embryos that had no chance of implanting in the uterus and that ordinarily would have been discarded. The separated blastomeres were coated with an artificial zona pellucida and placed in the culture medium used for in vitro fertilization (IVF).

The researchers obtained forty-eight blastomeres from the seventeen polyspermic embryos (eight two-cell, two three-cell, five four-cell, and two eight-cell), or theoretically forty-eight new totipotent embryos. A similar percentage of embryos cleaved for each stage of the embryo from which they were taken. While morulas (thirty-two-celled embryos) were achieved when blastomeres from two-celled embryos were cultured, blastomeres from four-celled embryos developed only to the sixteen-cell state, and no blastomeres derived from the eight-cell stage grew past eight cells in culture. These results suggest that splitting embryos at the two-cell stage appears to be more conducive to further development than does separation at the four-cell or eight-cell stage. However, the maximum stage at which a single blastomere can be reprogrammed to exhibit totipotency by itself or with cellular materials transplanted from other cells is unknown.[2]

The study thus demonstrated that experimental cloning or twinning of human embryos is potentially feasible as an aid to relieving infertility, though much additional work remains before offspring are produced, and there is uncertainty whether the technique will ever work at all. To produce a child by this method would first require showing that excised blastomeres from normal embryos would grow in culture to the point at which transfer to the uterus would ordinarily occur. Such research should also show the optimal stage for splitting normal embryos. It would then be necessary to place embryos that appear to be developing normally from split blastomeres into the uterus to show their potential for implantation and a successful pregnancy.

Some experts, however, are dubious that infertile couples would ever benefit from cloning by blastomere separation.[3] They view the higher pregnancy rate after transfer of several embryos as due to the genetic heterogeneity of the embryos transferred, not to numbers alone. On this view, placing several genetically identical embryos in the uterus will not increase the chances of pregnancy if one embryo with that genome would not have implanted. If this view is correct, there will be little incentive to use blastomere separation to treat infertility, and the ethical issues discussed below will have little practical significance. The following discussion, however, assumes that blastomere separation could provide certain advantages in treating infertility, and examines the ethical and policy issues that then arise.

Fears and Concerns

Some commentators saw nothing particularly unethical or disturbing in the George Washington research. This was simply another step toward improving the efficacy and efficiency of IVF; particularly for those couples who produce too few eggs or embryos to initiate pregnancy.

Many news reports, however, highlighted the disturbing or possibly unethical features of cloning and quoted ethicists who found the practice troubling. They described hypothetical scenarios in which embryos would be cloned for sale or to produce organs and tissue for existing children who needed transplants. One ethicist termed cloning as "contrary to human values"; others saw it as "an opportunity for mischief" that called for "governmental and societal debate and, perhaps, prohibitions and restraints."[4] The Vatican newspaper termed it a step into "a tunnel of madness," while the United Methodist Church called for an executive order banning cloning in all federally financed institutions.[5] A poll a week after the first story reported that 60 percent of Americans opposed cloning.[6]

The fears and concerns about cloning have several strands. Some of them arise from the artificial nature of assisted laboratory reproduction. Others are tied to discomfort with the manipulation and destruction of embryos that cloning research, if not the procedure itself, will inevitably cause. The most prevalent ethical concern, however, arises from the dangers that intentional creation of identical twins or multiples of one genome might pose to resulting offspring. The fear is that cloning will violate the inherent uniqueness and dignity of individuals, as well as create unrealistic parental expectations for their children. It also opens the door to identical embryos being created and sold because of their genetic desirability, as cattle embryos now are sold to increase animal yield and profitability. A worst-case scenario envisages the mass production of identical embryos to be sold to persons seeking desirable children. Finally, there are fears that embryos will be created to provide organs and tissue for existing children who need transplants.

Despite these reservations, research into the feasibility of splitting embryos will undoubtedly continue. Cloning by blastomere separation is basically a mechanical procedure that requires only the ability to micro-manipulate fertilized eggs and embryos and a few hundred dollars' worth of culture medium.[7] No DNA analysis or genetic expertise is necessary. It is likely that the next research steps—separating and culturing single blastomeres from normal embryos and then placing those that grow well into the uterus—and the actual birth of children as a result of embryo splitting might well occur in the next two to five years. As micro-manipulation of eggs and embryos is a rapidly growing practice, the ability to excise blastomeres from an embryo will easily be within the reach of many IVF physicians and embryologists. If shown to be safe and effective, physicians in many fertility centers will then offer the procedure to patients.

These possibilities engender a recurring disquietude about new reproductive technologies. Scientific zeal and the profit motive combine with the desire of infertile couples for biologic offspring to create an enormous power to manipulate the earliest stages of human life in infertility centers across the country. Even before one innovation is fully assimilated, the largely unregulated billion-dollar infertility industry presents another improvement, which separately or together threatens disturbing consequences for offspring, families, and society.

Some persons would argue that the idea of creating exact replicas of other human beings is so novel that there should be a moratorium on further research and development until a national consultative body evaluates the ethical acceptability of the procedure and develops guidelines for research and use of the technique. At the very least, to prevent abuses there should be strict rules about the circumstances in which cloning by embryo splitting occurs, and about the uses made of cloned embryos.

A closer look at the issues, however, suggests that the most likely uses of cloning are neither so harmful nor so novel that all research and development should now stop until the ethics of the practice are fully aired, or that governmental restrictions on cloning research or applications are needed. Indeed, there may be no particular need for guidelines beyond the full and accurate disclosure of risks and success rates that should always occur in assisted reproduction.

To assess the ethics of embryo splitting and the need for regulation, we must first ask who would use this technique if it were available and why, and then analyze the ethical issues that the likely demand for cloning would generate. We can then address the need for regulation of the embryo research that is essential if cloning by blastomere separation is to occur, and of the uses to which cloning technique will be put.

The Demand for Cloning

The news accounts of the George Washington University research emphasized many speculative uses of cloning, thereby slighting the most likely uses of the technique. The immediate impetus to develop cloning—and its most likely future use—is to enable infertile couples going through IVF to have a child.

To Increase the Number of Embryos Transferred

Initially the main demand for embryo splitting would come from couples undergoing IVF who cannot produce enough viable embryos to initiate pregnancy. In basic IVF practice, the highest rates of pregnancy occur with transfer of three to four embryos. Often more than that number of eggs has to be fertilized to produce enough viable embryos for transfer, with the excess frozen for use during a later cycle. Couples who produce only one or two embryos may thus have undergone an expensive and, for the woman, onerous procedure that has little chance of success.

Cloning by blastomere separation appears to be a reasonable step for such couples, if genetic heterogeneity of transferred embryos turns out not to be a key determinant of pregnancy success rates. Their goal is the birth of at least one child. If the prospective parents produce only two embryos, they would face the difficult choice of transferring those two in the hopes that a single pregnancy would result, or increasing their chances of having one child by splitting the blastomeres of one or both embryos.

If they produce only one embryo and embryo splitting has been shown to be safe and effective, they may opt to divide that embryo. Depending on the embryonic stage at which splitting is most successful, this could produce two embryos (if split at the two-cell stage), four (if split at the four-cell stage), or even eight (if two embryos are both split at the four-cell stage).

The number of embryos they end up with will affect the number of embryos placed in the uterus at any one time, and also whether cloned embryos remain available for transfer on a later cycle. If their efforts yield only two embryos, it is likely that both will be transferred to the uterus. (If both implant and come to term, embryo splitting will have produced identical twins).

If they produce four or more viable embryos by blastomere separation, three or four might then be transferred to the uterus in the hopes of having one child, with the rest frozen for later use.[8] Assuming two cycles of transfer with two to four embryos transferred in a given cycle, several possibilities arise. No children could be born from transfer in either cycle, or one or two could be born from the first transfer, and none from the second, or vice versa. In any given case, no child, one child, or deliberately created twins would have been born as a result of blastomere separation.

However, this scenario also opens the door to having "twins" (or even "triplets" or "quadruplets") born several years apart. This would occur if one or two children were born as a result of the first transfer cycle. Three years later, perhaps, the couple wishes to have a second child, and rather than go through IVF again, opts to have the remaining cloned embryos thawed and transferred to the wife's uterus. The period between births of children with the same genome could vary from a year or two to several years.

Embryo Splitting to Avoid Subsequent Egg Retrieval

Other scenarios involving embryo splitting as a treatment for infertility can also be envisaged. Perhaps the next most likely scenario if cloning by blastomere separation is in fact effective would arise with a couple undergoing IVF who produce a sufficient complement of viable embryos to initiate a pregnancy—three or four—but who wish to avoid the expense and burdens of subsequent egg retrieval cycles. Not many IVF candidates are likely to find themselves in this position, since ovarian stimulation often produces ten or more eggs. Because the couple would need to split only one or two of the three or four viable embryos that they have produced, it is conceivable that many couples who produce only four embryos would opt for this procedure. Indeed, the demand for embryo splitting from this group might arise even if it turned out that successful implantation requires genetic heterogeneity of embryos and the procedure thus was not sought by the group that produces very few eggs.

If some of their embryos are split but only noncloned embryos are transferred during the first or subsequent cycles, couples may satisfy their need for offspring without having to resort to cloned embryos. However, if uncloned embryos do not produce (enough) children, some of the cloned embryos may be thawed and transferred during a later cycle. In that case deliberately created twins could result at the same time, or at a point separated in time from the first child born with that genome. A third or fourth cycle using cloned embryos could result, with genetic replicas of earlier children born separated in time.

In either scenario, cloned embryos that are no longer needed by the couple that produced them might be discarded or donated to other infertile couples. Twins or triplets of an existing child might then be born to and reared by another couple. Because embryo donation is ordinarily anonymous, neither the children nor the genetic or rearing parents are likely to know the identity of the others.

Embryo Splitting as a Form of Life or Health Insurance

An often cited though highly unlikely demand for embryo cloning could arise from couples seeking insurance against disaster for any children that they have. That is, a couple might request that one or more blastomeres be split from embryos that will be transferred, so that the resulting clone can be frozen for later use in case the child born from the source embryo later dies or needs an organ or tissue transplant. In that case, embryos that are genetically identical to the child already born can be thawed and implanted in the mother (or a surrogate) to produce a genetically identical child to replace the dead child, or to serve as an organ or tissue donor for an existing child.

This scenario could occur, but it is unlikely for several reasons. First, few couples not otherwise undergoing IVF would choose to do so just to gain the hypothetical protection that identical backup embryos might provide. Second, couples that experience the death of a child may not, because of the sadness that it will engender, want to replace that child with a genetic twin, much less plan even before the first child is born to create a replica for that purpose. Third, couples undergoing IVF who produce enough embryos for transfer may not want to risk their viability by separating blastomeres for hypothetical insurance purposes. Fourth, a genetic replica of an existing child might not be necessary to provide needed organs or tissue, or there may not be sufficient time once organ failure in a child occurs to thaw, implant, and bring to term the cloned embryo to serve as an organ or tissue donor. Fifth, there may be medical reasons why a genetic twin will not be suitable as a donor, though in some cases, such as bone marrow or kidney transplantation, genetic homogeneity could provide an advantage.

Because so few couples—even those otherwise going through IVF—will request embryo splitting for this purpose, the use of cloned embryos as backup protection for existing children is likely to arise only with embryos that were created to enhance the efficiency of IVF. In situations of this kind, where the embryonic clones were *not* produced with the specific intention of insuring against disaster, parents might occasionally be glad of the opportunity to avail themselves of the stored cloned embryos to obtain tissue for transplant for an existing child, or to replace a child who has already died. Such scenarios are not impossible, but for the reasons stated above, they are not likely to be frequent.

Embryo Splitting to Obtain a Desirable Genome

Ethicists have speculated that cloning by embryo splitting might occur to facilitate, or might result in, the selection of stored embryos deemed to be particularly desirable. They envisage scenarios whereby parents will try to sell clones of desirable children to other couples, or where an attractive or successful couple will clone many embryos for later sale or dissemination.

These speculations are highly fanciful. Most couples are not in the market for other peoples' genetic offspring, but prefer to have their own. If so, they can exercise some control over the genetic characteristics of offspring by mate or gamete selection, or by preimplantation or prenatal genetic analysis. Few couples who can have their own children would be so obsessed with having a perfect child that they would eschew their own reproduction in order to obtain a cloned embryo that appears to have a desirable genome.

Of course, if cloning by embryo splitting is perfected, one could routinely excise and store a cell from every embryo that is produced and transferred to the uterus (assuming that this will not impair the embryo's development). The children born of the source embryo could then be followed, and the excised cells of those that turn out to have good genomes or healthy lives might then be sought by persons in quest of donor embryos. The mere description of the procedure shows how complicated and unwieldy it would be as a means to produce particularly desirable embryos.

However, couples who cannot produce genetic offspring might wish to have some say in the characteristics of embryos donated to them. In addition to choice of hair and eye color, and assurances that there are no genetic defects, they might want to see what the embryo they choose would look like as a child or youth, if such information were available. But there is no particular reason why it would be available, or why it would necessarily have to be provided.

In any event, providing information about cloned embryos to prospective recipients would not itself lead to embryo splitting specifically for purposes of genetic selection. The couple undergoing IVF might clone to enhance IVF efficiency, but there would be no particular point in cloning embryos just to enable genetic selection of donor embryos to occur at some later time. If the sale of embryos is also prohibited, the financial incentives necessary to induce embryo splitting for later sale would not exist.

Ethical Issues: Destruction of Embryos

Cloning by blastomere separation raises a number of ethical issues. Some ethical concerns derive from the stark interference with natural reproduction, or the manipulation and destruction of embryos that cloning necessarily entails. However, those concerns are not unique to cloning, and have been voiced about embryo research, freezing, and discard, and about IVF generally. Since they are not deemed sufficient to justify banning or restricting those accepted forms of assisted reproduction, they should not be sufficient to ban cloning either.

Yet persons who believe that fertilized eggs and embryos are already persons with rights will object that embryo splitting goes beyond the manipulations ordinarily involved in IVF. In this case a new unique individual will be intentionally split to serve other ends. The very process of blastomere separation could destroy embryos that would have developed normally, thus denigrating and undermining the value of human life. Because human life at all stages is a preeminent value, cloning by blastomere separation is an unethical procedure that should be banned.

There may be no way to answer the objections of persons who think that embryos are themselves persons and must be protected at all costs. The fact that embryo cloning might yield additional human lives will not assuage their concerns, for one is ordinarily not justified in killing one person in order to save several.[9] One can only point to the prevailing moral and legal consensus that views early embryos as too rudimentary in neurological development to have interests or rights.[10] On this view, splitting embryos can no more harm them than freezing or discarding them can. Nor is splitting embryos to enable one or more of them to implant and come to term inherently degrading or disrespectful of human life. Cloning embryos thus poses no greater harm to embryos than other IVF practices and should be permitted to the same extent that they are.

Ethical Issues: Deliberate Twinning

Ethical objections that are unique to cloning arise from a concern that the intentional creation of genetic replicas of an existing person denies the uniqueness of resulting offspring. This could occur from causing more than one child with the same genome to be born simultaneously. It could also occur from causing more than one child with the same genome to be born at different points in time.

Is the intentional creation of twins who are born simultaneously morally objectionable? Identical twinning occurs naturally and is not generally thought to be harmful or disadvantageous to twins. If anything, being a twin appears to create close emotional bonds that confer special advantages. If this is true, then having twins as a result of embryo splitting should be no more harmful to offspring than having twins naturally.

Suppose, however, that having twins does sometimes pose rearing problems or even psychological conflicts for children. For example, some families may have trouble rearing two infants simultaneously. Or asymmetrical relations with parents or intense rivalry between twins may occur, resulting in psychological harm to one or both of the pair. Still, the fact that undesirable outcomes might occur for some twins is no basis for concluding that all embryo splitting is unethical and should be discouraged.

The greatest chance that twins would result from embryo splitting would arise with a couple who produce too few embryos to have a reasonable chance of establishing even a single pregnancy. Their goal in embryo splitting is to have one child (or sometimes possibly two), but they know there is the risk that a twin pregnancy will result. If they knowingly accept the risks of twins, they will most likely be in a good position to handle the special burdens posed in rearing them. In any event, the risk of psychological harm from being a twin is neither so likely nor so severe that merely being born in this situation could constitute a harm. Twinning, whether natural or intentional, hardly amounts to a wrongful life. Neither child can reasonably claim that she has been wronged because, but for her parents' choice, she would have been born without a twin.

What, however, if triplets or even quadruplets are born simultaneously as a result of cloning by blastomere separation? If a four-cell embryo is split into four, and all separated blastomeres grow in culture and then are placed simultaneously in the uterus, the risk of a multifetal pregnancy increases. Multiple gestation does pose physical risks to the mother and to fetuses. Thus women who have multifetal pregnancies as a result of IVF or fertility drugs often use selective abortion to reduce the pregnancy to twins or triplets to improve the chances of a healthy outcome for all concerned. If multifetal pregnancies due to cloning occur, it is likely that they will also be selectively reduced to protect the health of mother and offspring.

Suppose a woman who is pregnant with triplets or quadruplets in virtue of transfer of four cloned embryos refuses to reduce the pregnancy to twins. Will she harm her offspring as a direct result of the cloning decision? Two different harms must be distinguished here. One is the potential harm of having three or four genetically identical siblings rather than just one, as occurs with twins. The second is the physical harm from prematurity that all offspring in such a multiple gestation might experience.

With regard to the first harm, it is not at all clear that identical triplets (or rarely, even quadruplets) suffer unique or inordinate psychological problems because they have identical siblings. If being an identical twin is generally a good thing, then it may be that being an identical triplet also has advantages and specialness that outweigh whatever disadvantages exist. At the very least, it would appear difficult to argue that these disadvantages are so great that the triplet should never have been born. Given that this is the only way for this individual to be born, its birth hardly appears to be a wrongful life that never should have occurred.

The risk of physical dangers of prematurity from a triple (or quadruple) pregnancy raises somewhat more complicated questions. Suppose the pregnancy ends prematurely at seven months. All three infants spend several weeks in intensive care and end up with permanent learning and physical disabilities. Have they been harmed by the cloning that produced a multifetal pregnancy which their mother refused to reduce to twins? The child who would have been aborted would not appear to have been wronged by the mother's refusal because it had no other way of being born but in a triplet situation subject to the very risks that have eventuated.

But two of the three infants (there is no way to identify the two that would not have been aborted) are worse off than they would have been if the pregnancy had been reduced to twins. Presumably they would then have been born healthy, without the physical and mental deficits they now have. One could reasonably argue that they have been hurt by their mother's refusal to reduce the pregnancy, even though there is no way to tell which two they are.

If the disabled offspring have been wronged, the wrong is not due to embryo splitting but rather to their mother's refusal to reduce the pregnancy from triplets to twins. The same arguable wrong would occur if the triple pregnancy occurred naturally or as a result of assisted reproductive techniques that did not involve cloning. Because the possibility of this wrong to the injured offspring is not unique to cloning, it is not an argument against all embryo splitting, any more than it is an argument against all use of fertility drugs or IVF, which also can produce multifetal pregnancies that are not reduced.

Ethical Issues: Later Born Twins

The second ethical issue unique to cloning by embryo splitting is the possibility of genetically identical siblings being born years apart in the same or different families. Are later born children harmed because a twin or triplet already exists? The claim rests on the notion that the later born child lacks the uniqueness or individuality that we deem essential to human worth and dignity, and that human individuality is largely determined by nature or genome rather than by nurture and environmental factors. Because phenotype and genotype do diverge, and because the environment in which the child will be raised will be different from that of his older twin, the child will still have a unique individuality. Physical characteristics alone do not define individuals, and there is no reason to think that personal identity will be wholly controlled by having an older twin.

Still, there could be special problems faced by such a child. Its path through life might be difficult if the later born child is seen merely as a replica of the first and is expected to develop and show the skills and traits of the first. This might be a special danger if the later born child is used as a replacement for an earlier born child who has died. However, it will be some years before the later born child is even aware of his genetic identity relative to an older sibling and the special expectations his parents might have.

But it is also as likely that the later born child will be loved and wanted for his own sake. His status as a later born twin (or triplet) could be seen as a special status, indeed, a unique or novel status that confers attention and love. It could also lead to close ties with the older twin, if the special bond that twins feel is genetically based. However, it could also lead to unique forms of sibling rivalry. Will the older twin feel that he is deficient because his parents wanted a newer version of him, or will he feel special and proud that his parents wanted another child like him? In any event, it is difficult to conclude that later or earlier born twins or triplets are likely to have such serious psychological problems that they should never be born at all. Even if one did so conclude, this would counsel against implanting cloned embryos only when a twin already exists, not against implanting two cloned embryos simultaneously or splitting embryos at all.

Ethical Issues: Cloning as Life or Health Insurance

Although cloning for the explicit purpose of providing parents with a replica for a lost child or as a source of organs or tissue for transplant for an earlier born child will not frequently occur, couples who have split embryos to treat infertility might occasionally be faced with thawing a cloned embryo for those purposes. Consider, for example, parents who request cloning to protect against the loss or death of a child, or who wish to thaw a cloned embryo to replace a dead child. Wanting a child to replace one who has died is not itself unethical. Nor does it become so merely because the new child will be a twin of the first. Although the parents may hope that the new child will develop and show the same traits as her deceased twin, they should very rapidly learn that the second child is different in some respects and similar in others, and would ordinarily come to treat and accept her as the individual that she is.

The use of cloned embryos as insurance against organ and tissue failure in an existing child presents a different set of issues. Here the concern is that the cloned embryo will be treated as an instrument or means to serve the needs of an older twin and will not be loved or respected for his own sake. As the Ayala case in California showed, however, a family can be motivated to have another child to provide an existing child with bone marrow and still treat the subsequent child with the love and respect that children deserve.

If this is so, thawing cloned embryos to provide tissue or organs for an existing child should also be ethically acceptable. The key is whether the child will be loved and accepted by the family that brings her into the world, not how or why she was conceived, nor even whether she was cloned for that purpose. As long as the child's interests are protected after birth occurs, it is hard to see how being cloned or thawed to provide organs for a twin is any worse than being conceived for that purpose. Even if it were, the risk that some cloned embryos might be used to provide tissue to existing children would not justify a ban on embryo splitting to treat infertility.

ETHICAL ISSUES: EMBRYO SPLITTING FOR GENETIC SELECTION

Scenarios involving embryo splitting for genetic selection are, as discussed above, extremely unlikely as long as overall demand for embryo donation is low and the buying and selling of embryos is not permitted. Since it is highly unlikely that a market in embryos will develop, there will be little incentive for couples going through IVF to clone embryos in order to sell them in the future. This is true even if recipients of donated embryos are permitted to pay some of the costs of embryo production.

It is true that the small subset of infertile couples who are candidates for embryo donation might wish to know the actual characteristics of existing twins or triplets of the embryos they seek to "adopt." However, neither having nor satisfying this wish is itself immoral. Indeed, the right of adoptive parents to receive as full information as possible about the children whom they seek to adopt is increasingly recognized. There is no reason why the same principle should not apply to embryo "adoptions." Even though the couple seeking the embryos will be choosing them on the basis of expected characteristics, such a choice is neither invalid nor immoral. As long as the parents are realistic about what the information signifies, do not have unrealistic expectations about the child's perfection, and love the child for itself, seeking and providing such information prior to embryo donation should be ethically acceptable. If it were not, providing such information could be banned without requiring that embryo splitting to treat infertility also be banned.

REGULATORY ISSUES

This account of cloning by embryo splitting and the ethical issues it poses suggests that, contrary to initial impressions, there is no major ethical barrier to proceeding with further research in embryo splitting as a treatment or adjunct to IVF. Given the great utility that embryo splitting could have for infertile couples, a moratorium on embryo splitting research is both unnecessary and unjustified. Such moratoria have occurred only when research appeared to pose great danger to others, as occurred with the brief moratorium on recombinant DNA research declared at Asilomar in 1975 because of the fear that genetically engineered pathogens might escape from the laboratory. By contrast, the risks of embryo splitting are no different from the risks that now exist in IVF laboratories and should be treated accordingly.

Even if a moratorium on cloning research is not justified, persons leery of embryo splitting argue for close regulation of the research that could perfect the practice, and then of its application. The most immediate policy questions concern whether there should be any restriction on research with embryos designed to improve or perfect techniques of embryo splitting. If research establishes the safety and efficacy of embryo splitting, then the question of regulatory limits on the use of the technique must be addressed.

The issue concerning the ethics of embryo research has several parts. One is whether research on normal embryos that will otherwise be discarded is ethically acceptable for any purpose. The second is whether embryos created by splitting blastomeres may ethically be placed in the uterus and brought to term. Such questions should be answered in terms of risks to the human subjects directly involved. As we have seen, the use of cloned embryos to treat infertility appears to be ethically acceptable. One should not deny investigators the right to carry out research otherwise respectful of human subjects because one disagrees with the utility or worth of eventual applications.

Research with Normal Embryos

The most immediate question is whether researchers should be permitted to split and culture blastomeres from normal embryos in order to replicate the results obtained at George Washington with polyspermic embryos. Although no authoritative American guidelines exist for research on normal embryos, there is a strong basis in the ethical literature and in practices of other countries for holding that such research is ethically acceptable.[11] Because early preimplantation embryos have no differentiated organs or nervous system, they cannot be harmed by splitting or other research manipulations and thus may ethically be used as the objects or vehicles of medical research.

As long as the research is for a valid scientific purpose, embryos that would otherwise be discarded can, with the informed consent of the couple whose gametes produced the embryos, ethically be used in research. Indeed, it should also be ethically acceptable to create embryos solely for research purposes when needed, even if there is no intent to place them in the uterus. Thus neither the lack of guidelines, the moral objections of some to any embryo research, or fears about where cloning research might lead justify forbidding researchers to take this next step. Researchers may not have the right to receive governmental or private funds for cloning, but if they are otherwise funded, their research should not be stopped because of objections to the use of embryos or to cloning itself.

Embryo Transfer after Splitting

Harder questions will arise if research shows that blastomeres separated from healthy embryos develop normally in culture to the point where they may implant in the uterus and go to term. Universities and IRBs might legitimately demand that implantation of manipulated embryos not occur until there are reasonable assurances that resulting offspring will not be physically harmed by the experiment. But if the embryos have developed normally in culture, this condition should be satisfied, just as it has been with embryos that were experimentally frozen and thawed before transfer, and with embryos that have been experimentally biopsied for preimplantation genetic analysis. When no physical harm appears likely, transfer to the uterus is ultimately beneficial for the resulting child (who has no other way to be born) and should be permitted. Again, neither the lack of clear guidelines, the fact that embryos will be manipulated and transferred, nor speculative fears of where blastomere separation ultimately could lead would be valid grounds for blocking this research.

Such a conclusion is consistent with the recommendations of commissions and advisory boards in the United States and abroad that have examined embryo research. Although they have not addressed cloning research directly, they do approve of transfer of embryos to the uterus after experimentation when the research is designed to aid or treat the resulting child. Transfer after experimental embryo splitting is designed to enable a child produced from blastomere separation to be born, and thus might be said to advance its interests. Just as the first embryo transfers after IVF were ethically acceptable because they enabled children to be born, so these should be as well, for there is no reason to think that if they implant and come to term they will have physical defects or otherwise be harmed.

Embryo Splitting Applications

Once it is shown that embryo splitting can produce normal offspring, the relative ease of the procedure and competition for patients will lead many IVF centers to offer it. Will it be necessary to restrict the uses to which embryo splitting is then put?

As the previous analysis suggests, the case for banning or greatly restricting embryo splitting as a treatment for infertility is extremely weak. The right of married and arguably even unmarried persons to procreate is a fundamental constitutional right that cannot be restricted unless clearly necessary to protect compelling state interests.[12] Because a ban on embryo splitting to treat infertility would directly interfere with the ability of infertile couples to have offspring, it would have to meet the compelling interest standard. Yet the prospect of great harm from intentional twinning, from twins born years apart, or from other possible uses of the technology does not appear to be so likely that governmental restrictions that go beyond assuring informed consent could be justified.

As with other forms of assisted reproduction, medical professionals who offer the service may be left largely to regulate themselves. IVF programs that engage in embryo splitting will have to decide at what stage embryos will be split, how many clones will be made, how many will be transferred at any one time, and how great a gap in time may occur between the birth of one child and another whose origin was the same embryo. They will have to develop procedures for counseling couples, particularly when twins are born months or years apart. Professional organizations, such as the American Fertility Society, might develop practice guidelines, as they have done with donor sperm and other reproductive technologies.[13] As long as the interests of couples and offspring are well served, there will be no need for governmental restrictions on the decisions made by medical professionals and their patients.

Nor do the more exotic scenarios imagined with cloned embryos necessarily warrant governmental intervention. The use of cloned embryos to replace a lost child or to provide tissue or organs for an existing child should be decided on the merits and ethics of those practices independently of creating or using cloned embryos for those purposes. If families may otherwise have children to serve as tissue donors for existing children, there is no basis for banning the use of cloned embryos for that purpose. Such uses are likely to be rare, and in any event, should not stop the use of cloning to treat infertility.

Similarly, couples seeking embryo donations should be entitled to as much information about the genetic characteristics of prospective offspring as is available. Wanting healthy, talented attractive children is not per se immoral and should not in itself bar the use of the technique. Of course if it did, that would not bar other uses of cloned embryos.

Laws that restricted trade or commerce in cloned embryos would be an acceptable public policy. Although it is highly unlikely that demand for cloned embryos would lead to a market in them, it maybe desirable to symbolize the unique value of incipient human life by banning the sale of embryos, whether cloned or not. Such a ban would not prevent infertile couples from getting access to infertility treatments or otherwise forming families, and thus would not limit or interfere with their procreative liberty. The ban need not prevent persons receiving embryo donations from sharing in some of the costs of embryo production.

The Permissibility of Cloning

The idea of cloning human beings initially sounds so bizarre and dangerous that one would think that such practices should be closely regulated, if permitted at all. Yet this survey of ethical and policy issues in cloning by embryo splitting suggests that the procedure has fewer risks and more benefits than first appeared and would be ethically permissible in most cases. The most unappealing applications of the technique are highly speculative and could be restricted without also stopping more valid uses.

Cloning by embryo splitting thus presents a regulatory situation that often arises with new reproductive technologies. An immediate step that seems justified to meet the legitimate needs of infertile couples could open the door to future applications that are much less defensible. If we ban the immediate steps in order to prevent potentially harmful future applications, infertile couples lose the benefits of the procedure without a clear showing that future harms would necessarily have occurred.

The temptation in such situations is to defer further research and development until a national commission or ethics advisory body puts its imprimatur on the practice. While such bodies, however, have been absent from bioethical debate in the United States for some time, there now appears to be an increased willingness to confront such issues. For example, an advisory panel on embryo research has been created to recommend guidelines for federal funding.[14] However, it remains uncertain when any such body will consider the complicated issues of human cloning.

As a result, we are left to elucidate and resolve on a retail basis the ethical dilemmas that each new innovation presents. Cloning by embryo splitting is another example of this policymaking process. Unless there are greater risks from its use than are now apparent, the case for adding the technique to the armamentarium of infertility treatments is a reasonable one. Its novelty will not prevent parents from loving and acting in the best interests of children born in this way.

Acknowledgments

The author gratefully acknowledges the comments of Howard Jones, Joe Massey, and George Annas on an earlier draft.

Notes

1. "Cloning Humans, *Time, 5* November 1993 (cover story); "Clone Hype," *Newsweek,* 8 November 1993, p. 60.
2. J. L. Hall et al., Experimental Cloning of Human Polyploid Embryos Using an Artificial Zona Pellucida," The American Fertility Society conjointly with the Canadian Fertility and Andrology Society, Program Supplement, 1993 Abstracts of the Scientific Oral and Poster Sessions, S1.
3. Howard Jones, Robert Edwards, and George Seidel, "On Attempts at Cloning in the Human," *Fertility and Sterility,* March 1994 (forthcoming).
4. Michael Waldholz, "Scientists Halt Research to Duplicate Human Embryos after Furor Erupts," *Wall Street Journal,* 27 October 1993; Gina Kolata, "Cloning Human Embryos: Debate Erupts over Ethics," *New York Times,* 26 October 1993.
5. *Time,* 5 November 1993; "Cleric Asks President for a Curb on Cloning," *New York Times,* 30 October 1993.
6. *New York Times,* 1 November 1993. Fifty-eight percent of respondents believed that "it was morally wrong to clone a human being." Sixty-three percent believed cloning was against God's will. Fewer than one in five respondents thought that cloning should be allowed to continue.
7. Statement of Dr. Robert Stillman on *Larry King Live,* 25 October 1993.
8. However, it is possible that three or even all four embryos transferred will implant. In that case, the couple will face the issue of selective reduction of the pregnancy to twins. Depending on the number of children who are born, cloning by separation could lead to twins or even triplets as a result of intentional cloning.
9. Judith Thomson, "The Trolley Problem," *Yale Law Journal* 94 (1985): 1395–1415.
10. American Fertility Society, "Ethical Considerations of the New Reproductive Technologies," special supplement, *Fertility and Sterility* 46 (1986); John A. Robertson, "In the Beginning: The Legal Status of the Early Embryo," *Virginia Law Review* 76 (1990): 437–517, at 440–50.
11. John A. Robertson, "Embryo Research," *Western Ontario Law Review* 24 (1986): 15–37.

12. John A. Robertson, "Embryos, Families, and Procreative Liberty: The Legal Structure of the New Reproduction," *Southern California Law Review* 59 (1986): 939–1041.
13. American Fertility Society, "Ethical Considerations of the New Reproductive Technologies."
14. Federal Register 59, no. 10 (14 January 1994): 2414. See also, Joseph Palca, "A Word to the Wise," in this issue (p. 5).

Splitting Embryos on the Slippery Slope: Ethics and Public Policy

by Ruth Macklin

Ruth Macklin is Professor of Philosophy and Medical Ethics in the Department of Epidemiology at Albert Einstein College of Medicine, New York City.

Neither the George Washington University embryo splitting experiment nor the technique of embryo splitting itself has ethical flaws. The experiment harmed or wronged no one, and the investigators followed intramural review procedures for the experiment, although some might fault them for failing to seek extramural consultation or for not waiting until national guidelines for research on preembryos were developed. Ethical objections to such cloning on the basis of possible loss of individuality, possible lessening of individual worth, and concern about potential harm to the resulting children are discussed and challenged, as are objections to the creation of embryos for the purpose of genetic diagnosis. Many of the ethical questions raised by the George Washington experiment are similar to those posed by existing reproductive technologies that allow the simultaneous production of several embryos. A multidisciplinary group should consider whether regulation of cloning is needed, and laws should be enacted to prohibit a commercial market for all frozen embryos.

Ruth Macklin, "Splitting Embryos on the Slippery Slope," *Kennedy Institute of Ethics Journal*, vol. 4, no. 3 (1994): 209–25.

IN OSCAR WILDE'S story, "The Picture of Dorian Gray," the character retains his youthful appearance in life while his portrait, painted when he was young, ages in the way a human being normally grows old. Periodic glimpses of his portrait reveal what Dorian Gray would look like if he—not the picture—bore the appearance of his chronological age.

In the furor that erupted following the experiment by scientists at George Washington University (Hall et al. 1993), ethicists envisaged possibilities that were termed, by analogy, "Dorian Gray scenarios." Arthur Caplan was quoted in the *New York Times* as saying: "Twins that become twins separated by years or decades let us see things about our future that we don't want to. . . . You may not want to know, at 40, what you will look like at 60. And parents should not be looking at a baby and seeing the infant 20 years later in an older sibling" (Kolata 1993b, p. C3).

A different assessment of identical twins born years apart was offered by Norman Fost. Fost was quoted as saying that although some people say it is chilling to think of identical twins born years apart, "it strikes me as better to have twins born years apart than to have them born together," since the duties of child-raising would be more spread out (Kolata 1993b, p. C3). And George Annas, debating the issue with Fost and others on "Charlie Rose," a late-night television talk show, said he wasn't sure whether two individuals who resulted from manipulating embryos in this manner but who were born years apart should even be called "twins."

The responses of these three prominent ethicists are remarkable more for the way in which separate concerns are jumbled together than for the variation in their assessments of the ethics of the experiment or its implications. Caplan's remarks focus on the psychological consequences for children who result from one split embryo, but are born years apart. Caplan may not want to know at 40 what he will look like at 60, but someone else might be eager to know that. In his comment about spreading out the duties of child-raising, Fost's observation appears to focus more on what is best for parents than for their children. Nevertheless, what is good for parents is often also good for their children. The lot of siblings might well be improved if their parents' duties toward them are spread over time rather than concentrated. On the other hand, if it is true that identical twins appear to have fewer psychological problems than other children (Paluszny and Abelson 1975), then the benefits to such children might outweigh the benefits to siblings whose parents had their child-raising duties spread over time. In questioning whether children resulting from one split embryo and born years apart should even be called "twins," Annas poses a conceptual question, one that should be kept separate from its ethical implications. The ethical issue addressed directly or indirectly in these comments is, however, central to the debate about embryo splitting: Are there likely or potential harms to the interests of children produced as a result of this technique? I shall return to this question later.

Possibly the most sobering lesson to be learned from the announcement that researchers at George Washington University had performed this research on a human embryo is that scientific misunderstanding can generate unwarranted fears, spontaneous overreactions, and focus attention on wild and improbable scenarios even among people who ought to know better. Different assessments of the success or failure of the experiment itself, as well as its ethical implications, sparked furious debates in the media and in living rooms throughout the nation and abroad. It is worth a brief look at these reactions and the ethical concerns people expressed before exploring whether the worries are justified.

The Experiment

The first set of questions pertains to the experiment itself: What did it accomplish? Was it truly ground-breaking? Was it an instance of the technique of cloning? Was the experiment itself unethical? And even if it was not intrinsically unethical, were the accompanying review procedures inadequate?

A statement by one of the investigators, Dr. Robert J. Stillman, released by the Office of Public Relations at George Washington University Medical Center, described the experiment as "a small step in an on-going scientific process" (Stillman 1993). Stillman noted that the study "involved the use of genetically abnormal embryos which developed from eggs that had been penetrated by more than one sperm and therefore could not have become viable." The stated goal of the experiment was eventually to "help infertile couples by reproducing nature's ability to reproduce twins." The press release closed on the cautionary note that "at the moment, this experiment has no clinical value to those whose histories might make them candidates for the potential procedure." According to Stillman's own acknowledgment, the experiment was not ground-breaking, but only "a small step in an on-going scientific process." Others questioned whether the George Washington experiment was at all original even in taking this small step, noting that similar research had already been conducted in numerous laboratories (Voelker 1994).

Was the experimental technique an instance of cloning? In referring to the Hall experiment as a "twinning technique," Jacques Cohen (1994) challenges the terminology that was widely used to describe the procedure. The disagreement over what to call the technique is not, as some might argue, "a matter of mere semantics." Had the experiment not been referred to as "cloning" in the popular media, no furor would have ensued. Jacques Cohen insists on a narrow definition of the term: "Cloning is taking the nucleus of a cell from the body of an adult and transferring it to an unfertilized egg, destroying the genome of the oocyte of the egg, and letting it develop" (Voelker 1994, p. 332). Stillman himself adopts a broader definition: "a clone is an identical replication, a duplicate copy" (Voelker 1994, p. 331). "I believe this is cloning," Stillman was quoted as saying (Voelker 1994, p. 331).

Were any features of the experiment itself unethical? As reported, there is nothing in the nature of the research itself that appears to have been unethical. No one was or could have been harmed by the experiment and no one was wronged. Although it is true that some people are opposed in principle to all embryo research, it is not a concern to be addressed here. The view that a preembryo has a moral status that would prohibit all *in vitro* or *in vivo* manipulations is part of a larger controversy over prenatal life, but not one that arises uniquely in the case of embryo splitting. A somewhat more permissive view would allow research on preembryos that could never become viable. The George Washington University experiment could meet this latter criterion since, as Stillman's press release stated, it "involved the use of genetically abnormal embryos which developed from eggs that had been penetrated by more than one sperm and therefore could not have become viable."

Could the George Washington University experiment be faulted on grounds of the stated purpose of this type of research? Although several ultimate purposes might be served by perfecting the technique of embryo splitting, the aim of these researchers was to "help infertile couples by reproducing nature's ability to reproduce twins." The researchers stated in their abstract that the technique "could be useful to patients who have difficulty producing a sufficient number of embryos for transfer" (Hall et al. 1993, p. S1). Even this rather limited goal could not have been accomplished in this particular experiment. Once perfected, the technique need not even involve gametes from a third party; it need not result in identical twins being born years apart; and it need not be done for any purpose other than increasing the number of embryos available to the infertile couple whose gametes produced the initial embryo. Of course, the ethical aspects of other possible purposes and consequences must be assessed, but they raise questions about safeguards, regulation, and prohibition that go beyond this experiment and its stated purposes.

Finally, there is the question of the review procedures surrounding the experiment. Did the researchers violate existing federal regulations? Were they remiss in not seeking additional layers of review before embarking on the experiment? Stillman reported that the experiment "had been approved by the university's formal review committees and peer reviewed by colleagues" (Sawyer 1993, p. A4). Furthermore, the research was carried out well within the 14-day limit for maintaining human preembryos for research endorsed by the Ethics Committee of The American Fertility Society (AFS) (1990), and it was in accordance with recommendations of the Ethics Advisory Board (1979), the Waller Commission (Committee to Consider 1984), and the Warnock Committee (Warnock 1984).

Nevertheless, some might fault the researchers for failing to take additional steps. In its report on research on preembryos, the AFS Ethics Committee (1990, p. 63S) stated that "[i]n some instances, the Institutional Review Board may well be advised to seek extramural consultation in order to broaden the moral judgment to be made," noting that "[t]he matter is of such grave public importance that approval of pre-embryo research should depend on conformity with guidelines established at the national level." There are, however, no such guidelines established at the national level or, for that matter, at state or local levels. The researchers might be faulted for not waiting until such guidelines were established. On the other hand, they surely could not fail to conform to national guidelines when such guidelines do not exist. The report of the AFS Ethics Committee is sufficiently vague in its wording that the researchers cannot be condemned for violation of unofficial guidelines for research on pre-embryos.

Responses in the Press

Responses recorded by journalists to the announcement of the experiment appeared immediately, but varied in their assessments of the degree of scientific advance and the ethical concerns posed by the research. An article in *Time* referred to the work as a "landmark experiment . . . different from anything that had preceded it" (Elmer-Dewitt 1993, p. 65). In contrast, Boyce Rensberger, writing in the *Washington Post,* said that "George Washington University's experiment at duplicating human embryos didn't work" since "the embryos mysteriously died very soon after cloning" (Rensberger 1993, p. A3). *Newsweek* accused the *New York Times* of an "apparent misunderstanding of a paper reporting a technical advance in embryology" (Adler, Hager, and Springen 1993, p. 60). The *Newsweek* article said that the *Times's* page-one story had a headline that "suggested that human embryos were being cloned in the laboratory." Presumably, the misunderstanding stems from the ambiguity of the term "cloning." According to Jacques Cohen's preferred definition, this was not cloning. According to the broader definition used by Stillman and others, it was.

The news accounts were, by and large, balanced assessments of the nature of the scientific advance; yet significant hyperbole could be found in some of these same articles with respect to the ethical concerns. The article that broke the story in the *New York Times* referred to "Brave New World" scenarios (Kolata 1993a) as did an article in *Science* (Kolberg 1993, pp. 652, 653). The article in *Time* claimed that "A line had been crossed. A Taboo broken. A Brave New World of cookie-cutter humans, baked and bred to order, seemed, if not just around the corner, then just over the horizon" (Elmer-Dewitt 1993, p. 65). The article quoted Jeremy Rifkin, founder of the Foundation on Economic Trends and a frequent critic of biotechnology, declaring "This is the dawn of the eugenics era" (p. 69). The *Newsweek* article included the subheading, "How Will the Clone Feel?" (Gelman and Springen 1993, p. 65).

In contrast to the modest assessments found in most of these articles regarding the scientific importance of the Hall experiment, comments like these in the popular press tend to create fears and misunderstandings among the public.

Responses by Bioethicists

Unfortunately, the reactions of bioethicists did not help to clarify the situation. Bioethicists' responses to any new biomedical or scientific development should properly take the form of reasoned analysis following careful study of the relevant facts and likely prospects. Few of us, however, can resist the temptation to reply to a journalist's phone call or a request to appear on a TV or radio show which offer the opportunity to utter only a sound bite. There is rarely an issue on which bioethicists are in complete agreement, and the embryo splitting experiment is no exception. What is, perhaps, surprising about the bioethicists' responses was the tendency to envisage or even contemplate worst-case scenarios. Margaret Somerville, from McGill University, was quoted as saying: ". . . what we are talking about is the ability to mass-produce humans" (Kolberg 1993, p. 652). The article in *Time* noted that "Ethicists called up nightmare visions of baby farming, of clones cannibalized for spare parts" (Elmer-Dewitt 1993, p. 65). Boyce Rensberger (1993, p. A3) in the *Washington Post* blamed ethicists for parlaying figments of their imagination into hypothetical scenarios, "bizarre scenarios of raising armies of clones or of creating twin siblings to harvest for organs."

When the bizarre scenarios are removed, the bioethicists' responses can he categorized into three broad groups: (1) expressions of opposition or serious ethical concern; (2) statements that there seems to be nothing ethically problematic about the experiment or what it portends; and (3) guarded statements about the need for further study, procedural safeguards, or additional layers of ethical review. Herewith, are examples in each category:

(1) Serious ethical concern:

—"The people doing this ought to contemplate splitting themselves in half and see how they like it" (Germain Grisez, quoted in Elmer-Dewitt 1993, p. 69).

—". . . we're in America, where we have the private market. We don't need government to make the nightmare scenario come true" (George Annas, quoted in Elmer-Dewitt 1993, p. 69).

(2) Nothing ethically problematic:

—It is the parents' prerogative to decide what to do with their embryos. [I start] "with a presumption of privacy and liberty, that people should be able to live their lives the way they want and to make babies the way they want" (Norman Fost, quoted in Kolata 1993b, p. C3).

(3) Guarded statements:

—"The first attempts to clone leave us with the possibility that we will create a lot of monstrosities along the way." But once the technique is perfected, "I don't see any reason why it is morally wrong" (Albert Jonsen, quoted in Kolata 1993b, C3).

—It is hard to argue that ethical principles would be violated by cloning; yet the technique could provide "an opportunity for mischief." That places a burden on those who would develop and offer the technique (Ruth Macklin, quoted in Kolata 1993b, p. C3).

Loss of Individual Uniqueness

Three prominent ethicists voiced concerns about the challenge to individual uniqueness raised by the Hall experiment:

—"One of the things we treasure about ourselves is our individuality. . . . You begin to worry that when you deliberately set out to make copies of something, you lessen its worth" (Arthur Caplan, quoted in Kolata 1993b, p. C3).

—"I think we have a right to our own individual genetic identity. . . . I think this could well violate that right" (Daniel Callahan, quoted in Elmer-Dewitt 1993, p. 68).

—"What do we mean when we talk about human identity? . . . How much of my identity is in having a genome from my parents that nobody else quite has? . . . We aren't just our genes, we're a whole collection of our experiences." But the idea raises a host of issues, "from the fantastic to the profound" (Albert Jonsen, quoted in Gelman and Springen 1993, p. 66).

What does the concern about individual uniqueness imply, and how are these comments to be understood? Caplan's remark focuses on the lessening of worth that follows from the deliberate production of copies of something. This focus addresses the counter charge that nature's identical twins lack genetic uniqueness and do not suffer from lesser worth because of that. So, presumably, if genetic duplication occurs in nature, there is no lessening of individual worth, but if done deliberately, it results in a lessening of worth.

But just what is meant by "lessening of worth," and to whom is the individual's worth lessened? Is it supposed that an individual will feel less worthy if he or she is the product of embryo splitting than if his or her existence comes about in some other way? It is an interesting psychological supposition, but one for which empirical evidence is entirely lacking. Is it supposed that individuals resulting from embryo splitting will be held by others to be less worthy? Less worthy than what? The analogy here is with works of art, such as lithographs, etchings, or woodcuts, produced by print techniques. The more prints that are pulled, the less each one is worth in monetary value. But surely monetary value is not the issue when contemplating the worth of cloned individuals to others. Two individuals resulting from a deliberately split embryo will have as much worth to their parents as identical twins that result from natural twinning. When they grow up, they will have as much worth to their friends, to their spouses, and to their children as do individuals who are genetically unique. Thus, the way in which deliberately cloned individuals have less individual worth remains puzzling.

Jonsen's remark (Gelman and Springen 1993) provides more insight but no answers. When he asks, "How much of my identity is in having a genome from my parents that nobody else quite has?" and replies that "We aren't just our genes, we're a whole collection of our experiences," Jonsen quite properly challenges the notion that genetics alone defines the uniqueness of each individual. This surely seems correct, but what are the implications for the relative weight of genetics versus experience in determining the worth of the individual? Jonsen's brief remark suggests an important distinction, that between *genetic* uniqueness and *individual* uniqueness. Even if the products of a split embryo lack genetic uniqueness, they nonetheless possess individual uniqueness, just like identical twins.

Caplan mentions only "individuality," thus blurring the distinction between individual uniqueness and genetic uniqueness. Callahan speaks of "individual genetic identity—"I think we have a right to our own individual genetic identity. . . . "—but oddly couches his remark in terms of rights. If this alleged "right" is violated by deliberate embryo splitting, why is it not also "violated" in nature when identical twins are born? Is "nature" somehow violating a right that individuals have? This phrasing is especially odd coming from Callahan, who has in many of his writings taken pains to question the proliferation of rights and the overuse of rights language in bioethics.

The earlier literature on cloning spawned by David M. Rorvik's 1978 book, *In His Image: The Cloning of a Man,* contained versions of this same ethical concern. The book, which purported to report genetic duplication of an adult human, was dismissed as a scientific hoax. However, it triggered the first wave of "clone furor," which included articles and ethical analyses by philosophers, scientists, and others. It should be noted that the form of cloning envisaged in those articles was not the embryo splitting or twinning technique, but the type of cloning that science fiction writers have depicted in stories about creating multiple copies of adult individuals and that conforms to Jacques Cohen's definition: "taking the nucleus of a cell from the body of an adult and transferring it to an unfertilized egg. . . ."

In the first wave of cloning literature, the argument about lessening the worth of genetically identical individuals also was dismissed by those who considered it. Martin LaBar (1984), for example, gives three replies. "First, it should be noted that the cytoplasm of eggs probably will affect development, so that cloned persons would resemble the person from whom their original nucleus was derived less than identical twins resemble each other (identical twins have arisen from the same original nucleus and cytoplasm)" (LaBar 1984, p. 325). It should be noted that LaBar is referring here to nuclear transplantation, so this objection would not pertain to the process of embryo splitting as performed by the George Washington University researchers.

"Second, identical twins have distinctively different personalities, according to common experience and to psychologists" (LaBar 1984, p. 325). This reply recalls Jonsen's observation that although two clones have the same genetic makeup they are whole collections of different experiences, which result in their being qualitatively different individuals.

LaBar's third reply, citing an article from the *American Journal of Psychiatry*, is that "identical twins appear to have fewer psychological problems than 'normal' children" (LaBar 1984, p. 325). This observation, however, properly belongs in a different ethical category. It does not respond to the concern about loss of individual uniqueness, but rather, to concerns about the psychological effects on children born as a result of cloning techniques. I shall turn to these concerns in a moment.

William A. W. Walters also considers and quickly dismisses the concern about individual uniqueness. Taking the notion to mean a "felt sense of uniqueness," Walters (1982, p. 112) writes:

> The latter [i.e. the cloned individual] might feel a lack of the sense of uniqueness or self when confronted by others of identical genotype. But . . . this argument fails to recognize that what distinguishes one human being from another is basically the unique pattern of roles and relationships he bears among his fellows and not any dissimilarity of his body from theirs.

Finally, a point made by Ruth Chadwick goes to the heart of Callahan's response. Chadwick (1982, p. 204) asks whether someone's *right* has been transgressed here:

> It is difficult to see how those moral philosophers who speak in terms of rights could make out a case for this. If everyone has a right to be genetically unique, then identical twins are cheated of that right. Further, in the cloning case, *who* is the possessor of the right? Not the person who is born as a result of cloning, because he would not have existed if he had not been cloned. There is no person who can be said to have the right to be genetically unique.

After considering both the arguments against cloning based on individuality concerns and replies to them, I conclude that either the objection itself is confused or incoherent, or else it should properly be subsumed under another category. The objection about loss of individual uniqueness is confused if it conflates genetic uniqueness with individual uniqueness. It is also confused if it contends that the worth of individuals is somehow lessened when genetically identical individuals are deliberately created, since the sense of "worth" involved in this concern views such individuals like multiple copies of an etching or lithograph. The objection is incoherent when cast in the language of rights, since there is no plausible candidate for an individual whose rights have been violated.

Finally, if the objection invokes the alleged psychological consequences for individuals who are cloned, it properly belongs in the category of harm to the interests of the child, to which I now turn.

Causing Harm to the Interests of the Child

"Suppose someone said, 'I've got 10 embryos stored, here's what they will become. You can look at the 20-year-old and see what you've got.' Is that fair to the child? What expectations will you put on them?" (Arthur Caplan, quoted in Kolata 1993b, p. C3).

This improbable scenario is only one of many examples cited of the potential harm to children who are products of embryo splitting or cloning. I can also be faulted for calling upon unlikely, but possible, future scenarios. "Suppose somebody wanted to advertise cloned embryos by showing pictures of already born children, like a product. . . . Unlike frozen embryos, you know how these are going to turn out. 'She's in first grade. See how cute she is'" (Ruth Macklin, quoted in Gelman and Springen 1993, pp. 65–66). More realistically, I noted that infertile couples are not very likely to give up a cloned embryo (Gelman and Springen 1993, p. 66).

It is often asked whether the entire array of new reproductive technologies is in the best interest of children. In the case of cloning, the concern is the potential harm to either an older or younger identical sibling brought into existence by this means. There is no way to predict accurately what psychological or emotional harm might result from this novel scenario. Parents generally want the best lives they are able to provide for their children, so one must wonder whether the nightmarish cloning scenarios that have been envisioned are likely to occur, for example, a couple deciding to replicate numerous children decades apart.

A more likely occurrence would be a decision to implant the second embryo from the original split two or three years later. This prospect recalls Fost's supposition mentioned at the outset, that is, the lot of siblings may be improved if their parents' duties toward them are spread out rather than concentrated. The most likely occurrence, however, is that of split embryos implanted simultaneously and born at the same time, in the manner of identical twins in nature. In this case, the finding that such twins have fewer psychological problems than other children becomes pertinent, and indicates that the children would benefit from, rather than be harmed by, the process.

But even if a couple were to choose to produce a cloned replica of an existing child, is it meaningful to ask whether the younger child would be harmed by being brought into existence? This is a child who would not otherwise be alive. How can we compare coming into the world as a cloned individual with never having existed at all? This line of argument has been used by John Robertson in several of his writings on the new reproductive technologies (1983,

1988, 1994). Ruth Chadwick (1982) makes essentially the same point in her criticism of the claim that in cloning, the "right" to genetic identity is violated. This consideration is a metaphysical argument rather than an ethical one based on consequences. It does not contend that more good than harm will result for children born as a result of embryo splitting or any other new reproductive technology. Rather, the argument questions the very meaningfulness of comparisons between never having been brought into being in the first place and the envisaged harms to a child brought into being in some ways rather than other ways. One reply to this argument takes the following form. The argument

> . . . justifies doing a wide variety of things that could be harmful to the child-to-be, as it can always be said that it is better to be alive, even though one suffers serious harms in being brought into existence, than not to exist at all. The point, though, is that there is no child-to-be waiting in the void of non-existence that could be harmed by not being brought into existence. The choice when embryo splitting is contemplated is between not having a child by means of this procedure, in which case there is no child that is harmed, and having a child by this procedure, in which case there is a child who, it is arguable, is harmed.[1]

This response is plausible in certain cases. For example, if a person is contemplating bringing a fetus with Tay-Sachs disease into the world, one can meaningfully ponder whether the pain, suffering, deterioration, and death of the child constitute a situation for the child that is worse than never having been born. The same question has been posed in the case of HIV positive women who decide to go through with a pregnancy. The HIV situation is much more difficult to assess because the birth of an afflicted child is only probable, 25 percent or less, not certain as in the case of a prenatally diagnosed Tay-Sachs fetus. In the case of children born with Tay-Sachs or HIV disease, the harmful consequences are predictable and serious. In contrast, however, the putative harms to children who are the product of embryo splitting are merely surmised, and the nature and severity of the envisioned harms are completely unknown.

I conclude, therefore, that ethical concerns about cloning that focus on potential harms to children born as a result of the procedure are, at best, based on imaginative suppositions. There exists little or no empirical evidence to support an ethical judgment that the production of children by embryo splitting would cause more harmful than beneficial consequences. As Robertson (1994, p. 10) notes: "Twinning, whether natural or intentional, hardly amounts to a wrongful life. Nor can one of them reasonably claim that they have been wronged because, but for their parents' choice, they would have been born without a twin."

Creating Embryos for the Purpose of Genetic Diagnosis

It is sometimes the case that an ethical evaluation of an act or practice rests on its purpose. Stillman described the purpose of the George Washington University experiment as eventually to "help infertile couples by reproducing nature's ability to reproduce twins." Women from whom very few eggs could be harvested for *in vitro* fertilization would then stand a better chance of achieving a pregnancy and birth. If embryo splitting were to be used for that purpose alone, an ethical evaluation would have to be based on the considerations discussed so far. But at least one other likely purpose—in contrast to the nightmarish scenarios—has been noted, namely, the splitting of embryos for the purpose of genetic diagnosis.

When this prospect was raised in connection with the George Washington University experiment, ethicists were again quick to comment. And once again, the responses range from guarded suspicion to moral objection to the view that there is nothing essentially wrong with the notion.

—The idea of "creating embryos solely for the purpose of genetic diagnosis is morally suspect" (Arthur Caplan, quoted in Kolberg 1993, p. 653).

—"You'd essentially have the situation of one identical twin being sacrificed for the sake of the other" (Margaret Somerville, quoted in Kolberg 1993, p. 653).

—The idea of using two- or three-cell clones for diagnosis is not much different from taking a single embryonic cell (John Robertson, quoted in Kolberg 1993, p. 653).

Caplan does not say why he finds the idea morally suspect. Is it for the reason Somerville gives, that one "identical twin" would be "sacrificed" for the sake of the other? Referring to these preembryos as "identical twins" is tendentious, as it conjures up the image of killing one Siamese twin to preserve the life of the other. In addition, the objection proves too much. It amounts to a much more sweeping condemnation of the destruction or disposal of all unimplanted embryos. It is self-deception to imagine that the 10,000 or more existing cryopreserved embryos will eventually be implanted. It is also folly to suppose that every time a couple decides to freeze three or four or seven embryos resulting from IVF they really intend to implant them at some time in the future. Except for those who are opposed in principle to the destruction of preembryos, it remains puzzling why the destruction of one preembryo for the purpose of determining whether its twin is free of genetic disease is ethically objectionable.

Robertson's contention is sound: The idea of using two- or three-cell clones for diagnosis is not much different from taking a single embryonic cell. It was recently reported that a healthy baby was born following the testing of the embryo for Tay-Sachs disease (Healthy Baby 1994). The preembryo was tested at the eight-cell stage along with three other preembryos from a couple who were carriers of the Tay-Sachs gene and had already had one child die of Tay-Sachs. A single cell was drawn from each of the four fertilized eggs. Three eggs were free of the genetic disease and were implanted in the woman. One healthy baby was eventually born. It is evident that some people who are opposed to abortion can nevertheless accept the destruction of an unimplanted preembryo. It is difficult to find a reason why anyone who finds abortion morally acceptable would find splitting an embryo for genetic diagnosis morally suspect.

Public Policy and Cloning

Before public policy discussions of cloning can begin, scientists must agree on a definition of cloning. Although this may appear to be a trivial point, it is conceptually necessary for distinguishing the perceived dangers of cloning from the actual implications of the Hall experiment, and from the different technique of nuclear transplantation. It is also necessary for providing an accurate picture of embryo research to the media, and thereby to the public, a requirement if public participation in debates of this sort is to be promoted.

Is the Hall experiment a case of science run amok? The answer depends on whether a long-overdue system of ethical review and oversight can be put in place for cloning and other new reproductive technologies. Although embryo twinning in the form carried out by the George Washington University investigators makes it possible deliberately to produce two or three individuals who may be identical, in most respects it does not pose ethical problems very different from those that attend already existing techniques. Current reproductive technologies enable a couple to produce numerous offspring simultaneously and to freeze the embryos and implant them at a later time. Children born in this manner are like fraternal twins that occur in nature, while children born as a result of the twinning technique are more like identical twins. It would indeed be novel for a couple to bear two or more identical children born years apart, but would it be ethically wrong? And if it would be, then what safeguards can be put in place to prevent these or other ethically unacceptable uses of the new technology?

The furor created by the cloning experiment has led to calls for legislation to ban further efforts of this type. Such a ban would be a hasty move, prompted by fear rather than by conclusions drawn from a thoughtful debate about what sort of regulation is needed. As Robertson (1994, p. 12) argues, "a moratorium on embryo splitting research is both unnecessary and unjustified. . . . [T]he risks of embryo splitting are not different than the risks that now exist in IVF laboratories, and should be treated accordingly."

What is needed by way of response to this scientific development is careful study, deliberation, and guidelines for the conduct of both research and clinical practice using preembryos. Careful study and deliberation might yield the conclusion that deliberate twinning is ethically acceptable for some purposes but not for others. Regulation is available as a public policy alternative to prohibition, but it first has to be clear just what should be regulated and for what reason. Although Stillman and other scientists described the George Washington University cloning experiment as merely "a small step in an on-going scientific process," some ethicists, other commentators, and members of the public saw it as a small step that begins an inevitable slide down a dangerous slippery slope. As in any such argument, the slipperiness of the slope, the dangers that lie ahead, and the possible means for stopping the slide must be explored sooner rather than later.

Although I have argued that neither the George Washington University experiment nor the technique of embryo splitting has intrinsic ethical flaws, the need to study the implications for other techniques and purposes remains. Careful exploration of such implications by a highly qualified, multidisciplinary group may yield the conclusion that some sort of regulation is needed. For example, once the technique is perfected it might be ethically prudent to limit the number of identical embryos that may be cloned from the original one. Some might urge setting a limit on the number of years split embryos may be kept in frozen storage in order to prevent women from implanting their own "clones"—identical "twins" who would be gestational mother and daughter. But a blue-ribbon panel is not needed to draw the conclusion that laws should be enacted to prohibit a commercial market not only for embryos produced by twinning or splitting, but for all frozen embryos. This would be one ethically desirable step in separating the commercial aspect of the new reproductive technologies from ethical concerns about their intended purposes and possible consequences.

Notes

1. This argument was formulated by Cynthia B. Cohen, former Executive Director of NABER, in a summary of issues related to cloning that was distributed to NABER members.

References

Adler, Jerry; Hager, Mary; and Springen, Karen. 1993. Clone Hype. *Newsweek* (8 November): 60–62.

Chadwick, Ruth F. 1982. Cloning. *Philosophy* 57: 201–9.

Cohen, Jacques, and Tomkin, Giles. 1994. The Science, Fiction, and Reality of Embryo Cloning. *Kennedy Institute of Ethics Journal* 4: 193–203.

Committee to Consider the Social, Ethical, and Legal Issues Arising from *In Vitro* Fertilization. 1984. *Report on the Disposition of Embryos Produced by In Vitro Fertilization.* Victoria, Australia.

Elmer-Dewitt, Philip. 1993. Cloning: Where Do We Draw the Line? *Time* (8 November): 65–70.

Ethics Advisory Board of the U.S. Department of Health, Education and Welfare. 1979. *Report and Conclusions: HEW Support of Research Involving In Vitro Fertilization and Embryo Transfer.* 4 May 1979. Washington, DC: U.S. Government Printing Office.

Ethics Committee of the American Fertility Society. 1990. Ethical Considerations of the New Reproductive Technologies. *Fertility and Sterility* 53 (Supplement 2): 62S–635S.

Gelman, David, and Springen, Karen. 1993. How Will the Clone Feel? *Newsweek* (8 November): 65–66.

Hall, J. L.; Engel, D.; Gindoff, P.R.; et al. 1993. Experimental Cloning of Human Polypoid Embryos Using an Artificial Zona Pellucida. The American Fertility Society conjointly with the Canadian Fertility and Andrology Society, Program Supplement, 1993 Abstracts of the Scientific Oral and Poster Sessions, Abstract 0-001, S1.

Healthy Baby Is Born After Test for Deadly Gene. 1994. *New York Times* (28 January): A17.

Kolata, Gina. 1993a. Scientist Clones Human Embryos, And Creates an Ethical Challenge. *New York Times* (24 October): A1, A22.

———. 1993b. Cloning Human Embryos: Debate Erupts Over Ethics. *New York Times* (26 October): A1, C3.

Kolberg, Rebecca. 1993. Human Embryo Cloning Reported. *Science* 262: 652–53.

LaBar, Martin. 1984. The Pros and Cons of Human Cloning. *Thought* 59: 319–33.

Macklin, Ruth. 1993. Cloning: Watch, but Don't Stop It. *Newsday* (5 November): 75.

Paluszny, Maria, and Abelson, A. Geoffrey. 1975. Twins in a Psychiatry Clinic. *American Journal of Psychiatry* 132: 434–36.

Rensberger, Boyce. 1993. The Frightful Invasion of the Body Doubles Will Have to Wait. *Washington Post* (1 November): A3.

Robertson, John A. 1983. Procreative Liberty and the Control of Conception, Pregnancy, and Childbirth. *Virginia Law Review* 69: 405–64.

———. 1988. Procreative Liberty, Embryos, and Collaborative Reproduction: A Legal Perspective. In *Embryos, Ethics, and Women's Rights: Exploring the New Reproductive Technologies,* ed. E. F. Baruch; A. F. Adamo, Jr.; and J. Seager; pp. 179–94. New York: Haworth Press.

———. 1994.The Question of Human Cloning. *Hastings Center Report* 24 (2): 6–14.

Rorvik, David. 1978. *In His Image: The Cloning of a Man.* Philadelphia: Lippincott.

Sawyer, Kathy. 1993. Researchers Clone Human Embryo Cells. *Washington Post* (25 October): A4.

Stillman, Robert J. 1993. Statement. (Undated press release from George Washington University Medical Center, Washington, DC.)

Voelker, Rebecca. 1994. A Clone by Any Other Name Is Still an Ethical Concern. *Journal of the American Medical Association* 271: 331–32.

Walters, William A. W. 1982. Cloning, Ectogenesis, and Hybrids: Things to Come? In *Test-Tube Babies,* ed. William A. W. Walters and Peter Singer, pp. 111–18. Melbourne: Oxford University Press.

Warnock, Mary. 1984. *A Question of Life. The Warnock Report on Human Fertilization and Embryology.* Oxford: Basil Blackwell.

Epilogue: Recent Developments

THERE ARE UNCONDITIONAL bans on human cloning in China, Australia, Israel, and most of Europe. In the United States, publicly funded research centers are prohibited from engaging in human cloning studies. Nevertheless, Dr. Richard Seed, a nuclear physicist and fertility specialist, made headline news in January 1998 when he announced plans to open a human cloning clinic in Chicago and declared that he would clone himself. In March 1999 he updated the plan, asserting that his wife Gloria had agreed to be the first human to be cloned and that she would be willing to bear her clone. Yet even such bold claims face competition, since, in the opinion of commentators, various privately funded laboratories in the United States are already vying with each other to produce the first human clone.

Cloning Human Embryos

Is human cloning far-fetched? In June 1999 researchers at the biotechnology company Advanced Cell Technology (ACT) in Worcester, Massachusetts, transferred the nucleus of a human cell from a man's leg to a cow's egg, which then developed into a human embryo. The ACT group claimed that this was the first time a human embryo had been successfully cloned. The embryo was destroyed within its first fourteen days, in order to comply with U.S. rules and to minimize criticism.

What is the purpose of cloning a human embryo? The research is carried out primarily with the intention of providing medical benefits for human patients with debilitating conditions. The idea is to extract from these early embryos stem cells, which are the "master cells" for all bodily tissue in that they possess the ability to differentiate into any tissue in the human body. In theory, having control over these stem cells means having the ability to grow any type of human body tissue under laboratory conditions. Stem-cell research is therefore expected to lead to the treatment of conditions such as diabetes and Parkinson's disease and to the formation of cells for organ transplant use.

ACT is not alone in this effort. Geron Corporation in Menlo Park, California, is also studying ways to clone human embryos. Having obtained rights to the cloning technology used by the Roslin Institute, Geron is now collaborating with Bio-Med, Roslin's commercial arm, in the attempt to clone human beings in order to create cellular spare parts for treating Alzheimer's, muscular dystrophy, strokes, and other debilitating conditions. This cloning research could lead to the production of human tissue that is totally immunocompatible. For example, it might provide heart muscle cells for heart attack victims, nerve cells for stroke victims and patients with Parkinson's and Alzheimer's, and liver cells to treat hepatitis, as well as tissues to treat degenerative disorders from cancer and spinal injuries. At this point, such medical advances seem to require cloning early-stage human embryos.

The subject of human cloning, however, has generated intense controversy. President Clinton banned federal funding for this type of research, but both ACT and Geron are privately held firms. Critics fear that human embryo cloning and stem cell research will eventually lead to whole-human cloning. And although companies like Geron have publicly opposed human cloning, opponents still point to an uncomfortable link between stem-cell research and reproductive cloning. Many of the opponents believe that early embryos are essentially human persons with full moral status, while supporters of the cloning process claim that personhood develops only after fourteen days and thereby justify use of the embryo before that time.

The debate over human cloning has taken on even more complex shades since scientists now point to a distinction between "therapeutic" and "reproductive" cloning. In therapeutic cloning, embryos are cloned in order to use their cells, especially the stem cells, for medical benefits, while in reproductive cloning, embryos are cloned in order to gestate as fully human babies. In the United States, although efforts to regulate private cloning research have not met with success in Congress, legislators are currently facing questions about the extent to which federal monies can be used for therapeutic cloning.

Cloning Animals

The rationale for therapeutic cloning is illustrated by the ongoing research into the cloning of animals, particularly livestock, which paves the way to the cloning of human embryos. Consider these highlights from recent years:

Early 1998: Research teams under the direction of Yukio Tsunoda successfully cloned cattle, a rare and costly commodity in Japan, where there are now herds of cloned cows. In only ten attempts, researchers were able to clone eight calves from the somatic cells of a single adult, and the group's 80-percent success rate was especially noteworthy, as it was the highest success rate so far reported. The commercial aspects of cloning cattle are far-reaching, both in terms of international trade policies and for the domestic Japanese market, where the strong demand for premium beef could be met with lower beef prices.

1998: Scientists at the University of Massachusetts at Amherst cloned a pair of cows, named Charlie and George, for use as a pharmaceutical source. Cloning such animals after altering their DNA is called "pharming," and the production and use of these transgenic animals has generated enormously lucrative investment opportunities. For instance, companies associated with the University of Wisconsin and Texas A&M University are aggressively investing in pharming.

April 1999: Scientists at the Oregon Regional Primate Center attempted to clone multiple rhesus monkeys in order to test new drugs for AIDS and some cancers. Their efforts evoked strong outcries from critics who feared that, because primates are genetically close to humans, this research would inevitably lead to human cloning.

April 1999: In a combined effort with teams from Tufts University School of Veterinary Medicine and Louisiana State University, researchers at Genzyme Transgenics Corporation in Massachusetts announced that they had successfully cloned three female goats from the same parent: Mira was born in October; the other two, also called Mira and Mira, were twins born in November. The goats were cloned specifically for the purpose of producing milk containing recombinant human antithrombin III (rhAT), a protein that helps to prevent human blood from clotting. Some of this rhAT has recently been tested on human patients; it has become the first transgenically produced protein to be used in clinical trials.

May 1999: Ian Wilmut and other researchers in Scotland publicly announced that the first cloned pig could be produced within months, at the latest by early 2000. This effort was sponsored by Geron Bio-Med, which intends to use pig organs for transplantation into humans. Up to this point, pig embryos had been cloned, but they had not yet survived in gestation. Two further obstacles to this xenotransplantation process still needed to be overcome: immunorejection and transmission of animal viruses into humans.

June 1999: Teruhiko Wakayama and Ryuzo Yanagimachi at the University of Hawaii produced the first male clone from adult cells, a mouse named Fibro. Prior to this, clones had been female, although scientists believe that certain male clones, such as prize bulls and endangered species, would be very valuable. Normal in all respects, Fibro was the single successful result from 274 attempts.

July 1999: In Scotland, the same researchers who cloned Dolly announced that they had successfully cloned two lambs, Cupid and Diana. What was unique about these lambs was that each was born carrying a particular gene that had been inserted by a new technique called gene targeting. The targeted gene carried by Cupid and Diana produces human serum albumin, which can be used to treat burn victims. Gene targeting involves replacing one gene with another in specific positions in the sheep's chromosomes; the technique can yield quicker production of human proteins in animals' milk. The research was undertaken by PPL Therapeutics in Edinburgh (the same company that funded the research on Dolly), which has already filed patent procedures for the technique.

September 1999: Scientists in Italy, under the direction of veterinarian Cesare Galli (who had worked with Ian Wilmut), announced the birth of Italy's first cloned bull, Galileo. In accordance with an official government ban on cloning, the Italian Health Ministry confiscated Galileo, which, like his namesake, is now serving time in a "pen."

September 1999: Two days later, the U.S. company Infigen, Inc., a unit of ABS Global in Wisconsin, announced that it had cloned a herd of thirty-seven cattle, probably the largest herd of clones in the world. Infigen, one of the leading biotechnology companies racing to clone farm animals in order to produce drugs for humans, had produced the first cloned calf, named Gene. Like Dolly, Gene was cloned from an adult somatic cell. Infigen is now collaborating both with Imutran, a unit of the Swiss company Novartis, to clone pigs for transplanting organs into humans, and with the Dutch firm Pharming Holding NV to produce cloned cattle whose milk can be used to treat human diseases.

Cloning Endangered Species

Although the cloning of animals is intended to benefit humans by treating diseases and providing viable organs for transplantation, some researchers claim that there are substantive benefits for animals as well, particularly animals that are extinct or in danger of becoming extinct.

In the bitter cold regions of Siberia, Frenchman Bernard Buigues has led efforts to unearth the two-ton, ten-foot-tall frozen wooly mammoth discovered there in 1997. Buigues hopes that the frozen mammoth, which is estimated to have lived between 8,000 and 15,000 years ago, may contain some perfectly preserved DNA that could be transferred to an elephant's egg and later gestated. Whether any DNA may be clonable is in question, and many scientists remain skeptical, especially since they contend that the entire genome of the mammoth needs to be intact, not just some isolated DNA.

In New South Wales, Australia, a government team is currently investigating the possibility of resurrecting the Tasmanian Tiger, an extinct species, by using the DNA from a specimen preserved since 1866. Although cloning this specimen successfully could eventually result in creating a viable gene pool, the project is estimated to cost more than $20 million, and many conservationists oppose the project on the grounds that the monies could be better spent.

In June 1999 researchers from the Chinese Academy of Sciences, under the guidance of Dr. Chen Dayuan, a zoologist, claimed to have successfully cloned the embryo of a giant panda by introducing cells from an adult panda into the egg cells of a Japanese white rabbit. The following October, the scientists announced their intention to implant the panda embryo into a black bear's womb in order to propagate the endangered species. Whether ongoing, aggressive measures like these will, in fact, save the panda from extinction is highly questionable.

Conclusion

In the face of criticism from conservationists and others, it seems that the more likely promise of cloning lies in the direction of cloning livestock. Even here, however, real dangers need to be addressed. To begin with, many cloned animals have developmental abnormalities as well as a high rate of neonatal death (especially among cloned sheep and calves), and they face the risk of premature aging. In May 1999 scientists at PPL Therapeutics in Scotland discovered that Dolly appears to be predisposed to premature aging. They base this claim upon studying her telomeres, nubs of nucleic acid and protein that protect the ends of chromosomes; the lengths of telomeres may account for longevity. Dolly has shorter telomeres than usual, probably either because her donor cell came from a six-year-old ewe or because her cell was cultured in the laboratory for some time.

Though the link between telomere length and longevity is still not certain, the real risk of premature aging remains, and such dangers make the notion of reproductive cloning all the more ominous. Critics of human cloning underscore these dangers, also asserting that the so-called therapeutic cloning of animals will become an inescapable bridge to the reproductive cloning of human beings. And since humans may be subjected to the same risks, research centers such as the Roslin Institute publicly oppose human cloning.

Will therapeutic cloning lead to the reproductive cloning of entire human beings? Is this a matter for regulation or is it simply a matter of time? The problem is all the more acute because, if it is true that companies are competing with each other to patent the human cloning process, there are obviously enormous financial implications, both in research costs and in the potentially gigantic profits to be reaped by the companies (and their shareholders) that do manage to obtain a patent.

The future hits us even before the present slips away. If human cloning is inevitable, and many—both critics and supporters—say that it is, does it hold out more promises or more perils? The stakes are high indeed, and for that reason, we cannot afford to wait passively for events to unfold. It is imperative that we continue to encourage and sustain reasoned discussion and debate in both professional and public forums.